Logic, Argumentation & Reasoning

Interdisciplinary Perspectives from the Humanities and Social Sciences

Volume 35

AF781112

Logic, Argumentation & Reasoning (LAR) explores links between the Humanities and Social Sciences, with theories (including decision and action theory) drawn from the cognitive sciences, economics, sociology, law, logic, and the philosophy of science.

Its main ambitions are to develop a theoretical framework that will encourage and enable interaction between disciplines, and to integrate the Humanities and Social Sciences around their main contributions to public life, using informed debate, lucid decision-making, and action based on reflection.

- Argumentation models and studies
- Communication, language and techniques of argumentation
- Reception of arguments, persuasion and the impact of power
- Diachronic transformations of argumentative practices

LAR is developed in partnership with the Maison Européenne des Sciences de l'Homme et de la Société (MESHS) at Nord - Pas de Calais and the UMR-STL: 8163 (CNRS) and the Center for Studies in Philosophy of Science, Logic and Epistemology (CEFILOE) - Instituto de Filosofía - Universidad de Valparaíso, Chile.

This book series is indexed in SCOPUS.

Proposals should include:

- A short synopsis of the work, or the introduction chapter
- The proposed Table of Contents
- The CV of the lead author(s)
- If available: one sample chapter

We aim to make a first decision within 1 month of submission. In case of a positive first decision, the work will be provisionally contracted—the final decision about publication will depend upon the result of an anonymous peer review of the complete manuscript.

The complete work is usually peer-reviewed within 3 months of submission.

LAR discourages the submission of manuscripts containing reprints of previously published material, and/or manuscripts that are less than 150 pages / 85,000 words.

For inquiries and proposal submissions, authors may contact the editor-in-chief, Shahid Rahman at: shahid.rahman@univ-lille.fr, or the managing editor, Juan Redmond, at: juan.redmond@uv.cl

Hykel Hosni • Jürgen Landes
Editors

Perspectives on Logics for Data-driven Reasoning

Springer

Editors
Hykel Hosni
Logic, Uncertainty, Computation and Information (LUCI) Lab, Department of Philosophy
University of Milan
Milan, Italy

Jürgen Landes
Munich Centre for Mathematical Philosophy
LMU Munich
Munich, Germany

ISSN 2214-9120 ISSN 2214-9139 (electronic)
Logic, Argumentation & Reasoning
ISBN 978-3-031-77894-0 ISBN 978-3-031-77892-6 (eBook)
https://doi.org/10.1007/978-3-031-77892-6

This Springer imprint is published by the registered company Springer Nature Switzerland AG
The registered company address is: Gewerbestrasse 11, 6330 Cham, Switzerland

Preface

During the first half of the past century, logic emerged as the field dealing with the methodology of deductive systems. Today, data-intensive and AI-driven science call for a new methodology of formalized inductive reasoning. The field is vast and currently in the making.

By soliciting the contributions appearing in this volume we aim to serve two sets of readers. First, we hope to stimulate logicians to apply their mindset and tools to the methodological and formal questions which arise within data-driven reasoning. Second, we hope to signal to the wider scientific community that logic can be vastly more useful in methodology than they expect. Arguably, the larger the intersection between those two sets of researchers, the more relevant logic will be in data-intensive and AI-driven science.

To make this volume particularly useful to early stage researchers, and then hopefully contribute to constructing a new research community centered on logic-based, data-driven reasoning, we have solicited contributions which put ideas before technical development.

Milan, Italy
Munich, Germany
January 2025

Hykel Hosni
Jürgen Landes

Acknowledgments

The initial idea for this book arose out of the Foundations, Applications & Theory of Inductive Logic project, which was generously funded by the German Research Foundation (grant number 432308570) and hosted by the Munich Center for Mathematical Philosophy. We would like to thank all the participants to this project, and in particular Soroush Rafiee Rad and Francesca Zaffora Blando who have encouraged us to pursue this project. Many thanks to both of you; this book has much profited from our discussions.

Obviously we are *very* thankful to the contributors, who were kind enough to accept our invitation to submit their work. We live in a time in which researchers of all seniorities are subject to enormous pressure to publish, with incentives usually not pointing at volumes like this one. This is a further reason for us to be sincerely grateful to the authors for being part of this project. We sent all submissions for peer review. Although they shall remain anonymous, we do want to extend our gratitude to the reviewers for their competent and very useful remarks which resulted in this being a much better volume.

We would also like to thank the team at Springer, who, with patience and competence, made this book a reality. Finally, many thanks to Shahid Rahman for thorough editorial work with unusual kindness, and to an anonymous reviewer for their constructive advice. Hykel Hosni acknowledges generous funding from the Italian Minister for University and Research (MUR) in the form of the FIS 2021 *Advanced Grant* "Reasoning with Data" (ReDa) G53C23000510001.

Jürgen Landes gratefully acknowledges NextGenerationEU funding for the project "Practical Reasoning for Human-Centred Artificial Intelligence" as well as funding from the Deutsche Forschungsgemeinschaft (DFG, German Research Foundation) under the grant 528031869.

Contents

Contributors

Paolo Baldi University of Salento, Department of Human Studies, Lecce, Italy

Esther Anna Corsi Logic, Uncertainty, Computation and Information (LUCI) Lab, Department of Philosophy, University of Milan, Milan, Italy

Maria Luisa Dalla Chiara Dipartimento di Lettere e Filosofia, Università di Firenze, Firenze, Italy

Francesco Facciuto Logic, Uncertainty, Computation and Information (LUCI) Lab, Department of Philosophy, University of Milan, Milan, Italy

Roberto Giuntini Dipartimento di Pedagogia, Psicologia, Filosofia, Università di Cagliari, Cagliari, Italy
Institute for Advanced Studies, Technical University Munich, Garching, Germany

Hykel Hosni Logic, Uncertainty, Computation and Information (LUCI) Lab, Department of Philosophy, University of Milan, Milan, Italy

Jürgen Landes Munich Center for Mathematical Philosophy, LMU Munich, Munich, Germany
Logic, Uncertainty, Computation and Information (LUCI) Lab, Department of Philosophy, University of Milan, Milan, Italy

Chiara Manganini Logic, Uncertainty, Computation and Information (LUCI) Lab, Department of Philosophy, University of Milan, Milan, Italy

Giuseppe Primiero Logic, Uncertainty, Computation and Information (LUCI) Lab, Department of Philosophy, University of Milan, Milan, Italy

Giuseppe Sergioli Dipartimento di Pedagogia, Psicologia, Filosofia, Universitá di Cagliari, Cagliari, Italy

Alena Vencovská The Open University, Milton Keynes, UK

Felix Weitkämper Institut für Informatik, Ludwig-Maximilians-Universität München, München, Germany

Jon Williamson Department of Philosophy, University of Manchester, Manchester, UK

Sandy Zabell Department of Statistics and Data Science, Northwestern University, Evanston, IL, USA

Acronyms

AF	Argumentation Framework
DAG	Directed Acyclic Graph
ERC	European Research Council
Ex	Constant Exchangeability
INV	Invariance Principle
IP	Constant Irrelevance Principle
KRR	Knowledge Representation and Reasoning
Li	Language Invariance
MAP	Maximum-a-Posteriori
ML	Machine Learning
NHST	Null Hypothesis Significance Testing
OBIL	Objective Bayesian Inductive Logic
OBN	Objective Bayesian Network
PAC	Probably Approximately Correct
PI	Principle of Induction
PIL	Pure Inductive Logic
PIP	Permutation Invariance Principle
PIT	Preservation of Inductive Tautologies
Px	Predicate Exchangeability
SN	Strong Negation Principle
StarAI	Statistical Relational Artificial Intelligence
Sx	Spectrum Exchangeability
UPI	Unary Principle of Induction
Vx	Variable Exchangeability
WIP	Weak Irrelevance Principle

Chapter 1
A Note on Logic and the Methodology of Data-Driven Science

Hykel Hosni and Jürgen Landes

Abstract This introductory, editorial chapter sets the stage for the following contributions by discussing roles that logic has—and has not played—in the development of reasoning under uncertainty tracing from Boole and De Morgan, over Tarski to AI, the AI spring, and current trends in data-intensive methodology.

Keywords Data-driven science · Logic · Tarski · Metamathematics · Data-driven reasoning · Artificial intelligence

1.1 Introduction

Data-intensive and AI-driven science is coming of age. Owing to an unprecedented ability to produce, gather, and analyse vast amounts of data, over the past decade, success has continued to come at an increasingly fast pace across all fields (Hey et al. 2020), reaching spectacular peaks in the life sciences (Ramsundar et al. 2019; Rodriguez et al. 2021; Senior et al. 2020; Sim 2016; Wang et al. 2023). While forecasting the evolution of hype-prone AI is notoriously tricky, we believe that data-intensive and AI-driven methods are here to stay, and that they will shape an increasingly larger proportion of research in the decades to come. But along with the promises of scientific "acceleration" a number of challenges arise, chiefly about the transparency, explainability and, ultimately, trustworthiness of this relatively new way of producing *and* transferring scientific knowledge to society (Nowotny 2021).

H. Hosni (✉)
Logic, Uncertainty, Computation and Information (LUCI) Lab, Department of Philosophy, University of Milan, Milan, Italy
e-mail: hykel.hosni@unimi.it

J. Landes
Munich Center for Mathematical Philosophy, LMU Munich, Munich, Germany, Milan, Italy

H. Hosni, J. Landes (eds.), *Perspectives on Logics for Data-driven Reasoning*, Logic, Argumentation & Reasoning 35,
https://doi.org/10.1007/978-3-031-77892-6_1

We believe that the logician's style of thinking has lots to contribute to making the most of the opportunities while mitigating the risks of this emerging way of doing science. Hence, the volume aims primarily at putting forward a state of the art presentation of the role logic can have in answering the call. And since all this is in the making, one key goal of the book is to motivate more logicians to engage actively with the data-centred turn in science which, as illustrated in the chapters to follow, is dotted with problems waiting to be tackled by the wider logical community.

But you may—and should—ask: *why are you trying to engage logicians? Aren't the people who know most about logic hard at work figuring out data-driven reasoning already? They surely must be excited about the prospects of the relevance of their work in shaping the new science!* We have no historical nor sociological data to offer in response to this key question, but we believe that an (openly opinionated) outline of the recent trajectory of the field can help pinning down the roots of the *status quo*.

1.2 The Restriction of Logic to Metamathematics

Alfred Tarski, as reported by his student John Corcoran, purportedly referred to himself as "the greatest living sane logician." This would allow Tarski to position himself above Kurt Gödel, widely recognized as the foremost logician of his time, yet whose mental state sadly fell short of sanity. Be this as it may, Tarski is certainly one of the great modern logicians. He is also the author of the first popular-science book in logic, which shaped the subject and its image within the wider scientific landscape. Written in Polish in 1936, it appeared in German with the title *Einführung in die mathematische Logik und in die Methodologie der Mathematik* and it was turned into a textbook for the 1941 English translation, titled *Introduction to Logic and to the Methodology of Deductive Sciences*. As explained in the Preface, Tarski contrasts the latter with "the empirical sciences" which do not lend themselves purposefully to logical methods:

> I see little rational justification for combining the discussion on logic and the methodology of empirical sciences in the same college course (Tarski 1941, p. xiv).

This is not to diminish the importance of empirical science (of course!). Indeed in a comment to a seminar quoted in Chapter 10 of Feferman and Feferman (2004) Tarski states:

> It would be more than desirable to have concrete examples of scientific theories (from the realm of the natural sciences) organized into deductive systems. Without such examples there is always the danger that the methodological investigation of these theories will, so to speak, hang in the air. Unfortunately, very few examples are known which would meet the standards of the present-day conception of deductive method and would be ripe for methodological investigations [...] The development of metamathematics, that is, the methodology of mathematics, would hardly have been possible if various branches of mathematics had not previously been organized into deductive systems.

This way of thinking, which puts forward axiomatisation as a necessary condition for the applicability of logic to scientific methodology, has been very influential in shaping the twentieth-century view of the subject, in and outside logic. It effectively means centring the applicability of logic to the foundations of mathematics—a view which persists to date, as certified by the current European Research Council (ERC) panel structure.

We take, of course, no issue with giving importance to this branch of logical research. But we think it is useful to remind our readers that, albeit very fruity, this branch is neither the tree, nor its root.

1.3 Choosing a Better Level of Analysis

To be clear, classical logic, i.e. the set of tools and techniques which originate with metamathematical investigations, provides sound-enough foundations for (virtually all of) classical mathematics, within which every correct proof can be given in terms of logical inferences. Since formal reasoning in the sciences is basically mathematical, *formal* scientific inference can—in principle—be done in classical logic. So, going beyond the classical view logic doesn't mean rejecting it, in principle. The qualifier "in principle" is the key point.

Consider the following analogy. In principle it is possible to use operations on binary strings to compute with real numbers. Take the still rather simple computation $5^{12} + 237$ and imagine reducing the exponentiation to multiplication:

$$5^{12} = 5 \cdot 5 \cdot 5 \cdot 5 \cdot 5 \cdot 5 \cdot 5 \cdot 5 \cdot 5 \cdot 5 \cdot 5 \cdot 5$$

the multiplications to additions:

$$5 \cdot 5 = 5 + 5 + 5 + 5 + 5$$

and the real numbers to binary numbers

$$5 = 101.$$

Not only is this type of computation painstakingly time-consuming, we—as human reasoners—fail to see the underlying structure of the computation. The interesting abstract features of the computation have disappeared. A computation of the term $5^{12} + 237$ at the usual level allows us instead to decipher the presence of an exponentiation, which is arguably a necessary condition to take exponentiation as an object of mathematical investigation. The general, and rather obvious, point here is that it is the *choice of the level of analysis*, which can facilitate insights or prevent successful inquiry. To close our detour, if we are to use logic in the methodology of data-driven science, the level of analysis offered by classical logic is hardly our best option.

This of course opens the question as to which is our best option, or even a good one to begin with. Since what counts as a good scientific inference changes over time, differs from discipline to discipline, and sometimes even from application to application (the philosopher of science Adrian Currie coined the suggestive term "methodological omnivory" applying to scientists who construct purpose-built epistemic tools tailored to generate evidence about highly specific targets Currie 2015), the prospects for providing *the* logic of data-driven inference are dim. The challenge thus arises to formulate and formalise frequently-used patterns of scientific inference as a guiding ideal of valid data-driven reasoning, while providing a formal inference apparatus that is sufficiently tractable to be useful. What exactly counts as "frequently-used", "tractable", and "useful" of course depends on the scientific problem at hand, the specific circumstances and the training of the working scientists.

1.4 AI Revamped the Boolean Roots of Logic

Luckily, but unknown to many scientists, the logical landscape has broadened significantly since Tarski's landmark *Introduction* and features many *non-classical logics* devised to elucidate practical inference, covering certain areas and aspects of scientific reasoning. In particular, moving out of the main stream of its glorious past, a relatively small but dedicated legion of logicians around the 1980s sought to apply their tools, techniques, and crucially their mentality, to questions arising within the then-emerging fields of computer science and artificial intelligence. This led to a reconciliation of logic with the psychology of reasoning, decision-making, and reasoning under uncertainty not seen since the pioneering work of George Boole and Augustus De Morgan which the metamathematical view of logic had relegated to temporary disgrace.

Indeed Boole and De Morgan, in the mid-nineteenth century, launched the ambitious plan of tackling logical reasoning with mathematical tools, thereby inventing *mathematical logic*—an expression apparently coined by De Morgan. And it is particularly noteworthy that both considered probability to fall within the scope of logic. De Morgan's 1847 book is titled *Formal Logic: Or the Calculus of Inference Necessary and Probable*, whereas Boole's 1854 follow up to his *Mathematical Analysis of Logic* is titled *An Investigation of the Laws of Thought on Which are Founded the Mathematical Theories of Logic and Probabilities*. To both of them, the object of logic is reasoning, and since reasoning often proceeds under uncertainty, probability should be developed within logic. Strikingly simple.

Since we are not historians, we will not venture into the explanation as to why this broader and ultimately more applicable view of logic was soon to be superseded by the formalist view of the Peano, Russell, Frege, and Hilbert variety. We only allow ourselves the liberty to note that whilst this view brought about among the most admirable theorems in all of mathematics, it had the side effect of restricting the scope of logic to metamathematics, as crystallised by Tarski's *Introduction*. It is not implausible that this had a strong impact on what two generations of logicians have considered a question worth asking. And restricting the logicians' attention to questions that could be answered metamathematically, effectively meant leaving the problems of interest in the methodology of data-driven inference largely unattended.

A major shock came with the AI spring which bloomed in the 1980s. The relative success of *expert systems* and the general interest in formalising *common sense reasoning* led the search for seamless connection between logic and probability. Several journals were started around that time to accommodate this rediscovery of the role of logic in (uncertain) reasoning, including the *International Journal of Approximate Reasoning* (1987); the *Journal of Logic and Computation* and the *Journal of Applied Non-Classical Logics* (both in 1991); the *Journal of Logic Language and Information* (1992); and the *Logic Journal of the Interest Group in Pure and Applied Logic* (1993). We refer to Denœux et al. (2020) for a comprehensive survey of the impact this movement had in AI in this, and to Hansson (2014) for a telling illustration of how "classical methods" have become crucial in investigating "non-classical problems" in logic.

Unfortunately, too few outsiders to the logic profession have noted how logic reclaimed some of its original breadth in the attempt to deliver the promises of computation-based intelligent behaviour. This can be partly explained by the specialisation and sheer volume of published research, and partly by the fact that popular-science books in logic still indulge non-specialist readers with accounts of the heights and depths reached by the golden age of metamathematics. Again, there is nothing wrong with this, except possibly the fact that "other scientists" have little opportunity to appreciate the variety of logics which have been developed to capture interesting aspects of reasoning and inference which are not restricted to mathematical proofs. And among those logics, many can be put to work in providing the urgently needed methodological scaffolding to data-intensive and AI-driven science.

1.5 Which Logic for Data-Driven Reasoning?

According to the twentieth century cliche, the core of the scientific method involves putting forward hypotheses, the deductive consequences of which are to be tested empirically. This is crystallised in the charming Feynman dictum to the effect that *if it doesn't agree with experiment, it's wrong*. However this is viable only in (certain consolidated areas of) physics, where the objects of reasoning and inference are essentially mathematical. Everywhere else the dictum is easy to state, but practically hard to apply. For the data featuring in scientific reasoning are often gappy, messy, if not outright contradictory. Moreover there is seldom a unique way of mathematising the concepts of interest, think for example of the concept of "rationality" which underpins the social sciences, or the identification of the relevant interactions in complex systems which permeate the life and natural sciences. Contemporary philosophers of science seem to agree that all of this shows that the cliche is way too simplistic (Cartwright et al. 2022; Lewens 2015). And yet without sound-enough reasoning there cannot be any science. So, if classical logic is not the right methodological tool for data-driven inferences, what is?

1.5.1 Don't Trust the Analogy Principle

Possibly owing to the fact it is the only logic they have some familiarity with, many scientists and methodologists of science seem to be comfortable with taking classical logic as a justification for the validity of the inferences they seek to make from data. When statisticians who support Null Hypothesis Significance Testing are questioned about its validity, they are likely to invoke an analogy with *modus tollens*: Infer that the null hypothesis H_0 is false from (1) the assumption that H_0 implies data D and (2) the observation D is extremely improbable. Many issues can be raised against this formulation of modus tollens (Wagner 2004), but one really stands out: D being *extremely improbable* and D being *false* are qualitatively distinct. And yet, many scientists don't seem to make much of this. A telling case in point is eminent biostatistician Richard Royall:

> [Modus tollens] is at the heart of the philosophy of science [...]. Its statistical manifestation is in the formulation of hypothesis testing that we will call 'rejection trials'. [...] the form of reasoning in the statistical version of the problem parallels that in deductive logic: if H_0 implies D (with high probability) then not-D justifies rejecting H_0". (Royall 1997, pp. 72–73)

This way of justifying the inferential conclusion of hypothesis testing, and more generally, the kind of reasoning underpinning the Feynman dictum, relies on the unstated assumption to the effect that data-driven patterns of inference borrow their validity from analogous classical patterns. Now, there is a problem with this *analogy principle*, and it is that it does not (always) work.

For a simple example, take George Polya's *fundamental inductive pattern* (Pólya 1968):

$$\frac{\begin{array}{r} A \rightarrow B \\ B \end{array}}{A \text{ [becomes] more plausible}} \tag{1.1}$$

As a scheme of inference, (1.1) is has no direct counterpart in classical logic, but if one reasons by analogy and takes "A [becomes] more plausible" as "A [is] true", then (1.1) becomes classically *invalid*. No wonder then that rather spectacular failures of the analogy principle can be observed in connection to the (mis)applications of hypothesis testing (Hofmann 2002; Krueger 2001; Patriota 2018).

The general lesson we can learn from the failure of the analogy principle, if one wants to call it like this, is that valid inference is not a continuum spanning mathematics and the empirical sciences. This certainly vindicates Tarski's insistence on classical logic having a direct and unmediated role in the methodology of deductive sciences *only*. However, one century on, both the scientific and the logical landscapes have changed dramatically. And since the penetration of AI through machine learning technology in science, institutions, and our own daily lives is pushing us hard to devise a fail-safe "general method" for data-driven inference, the time is ripe for logicians to venture where Tarski saw no role for logic.

1.5.2 A Broader Understanding of Logic

The first difficulty encountered in this exploration is to do with the very meaning of logic. In traditional mathematical logic the syntax, semantics, proof-theory and model-theory of any logical system are defined rigorously. Other than that, logicians know a logic when they see one. This is hardly sufficient to pin down precisely the features that a piece of formalisation should satisfy for it to be called a logical system (Gabbay 1995).

So we certainly do not attempt at a definition here. Rather, we embrace the fertile ambiguity of the word *logic*. In addition to being identified with metamathematics, logic can also be taken informally as a description of the inner working of things. In a long thread which includes Maxwell as an early contributor, many scientists incorporated the centrality of uncertainty in scientific reasoning by identifying "the logic of science" with probability theory, a role which in the age of data-intensive and AI-driven science is perhaps best attributed to (inferential) statistics.

In many ways the state of the art in the field is comparable to that of the early 1800s, prior, that is, to the revolutions brought about by George Boole. Before his "general method" logicians used to discuss individual patterns of reasoning—syllogisms. It is perhaps no coincidence that a rather common way of referring to the inference underlying NHST as *statistical syllogism* (Chow 1997).

This is precisely why we think the problem of identifying criteria of validity for data-intensive and AI-driven inference should be of interest to logicians, in addition of course, to the methodologists of science. The contributions below certainly do not provide a unique or general answer to this problem. In fact, some of them do not even look like they are about logic, if this is what we expect the logicians who only focus on metamathematics to think about it. But we solicited them because we think they contribute towards formulating central desiderata for what the *Logic for Data* should be.

1.5.3 A Virtuous Circle

Whilst it is possible and fruitful to tackle the question from the metamathematical perspective, see e.g. Ben-David et al. (2019), our present focus is broader. To make a pictorial analogy, our view embodies the kind of virtuous circularity depicted in Cornelius Escher's iconic *Drawing Hands*. The one hand is the data-driven AI spring, and we ask how this impacts on the other hand, the way logic is done. And of course we ask the converse question as well: how can the thus-emerging logic contribute in the methodology of data-intensive and AI-driven science, along with its many societal ramifications.

The virtuous circle joining logic and *technology* can be found, again, at the Boolean roots of contemporary logic. The idea that reasoning could be mechanised is in the background of many contributors at least since al-Khwārizmī's *Algebra*, circa 830. However the visible impact of the machines driving the industrial revolution in Victorian England is likely to have given this idea flesh and bones. This is how historians put it:

> The nineteenth century was the first time in human history when large numbers of ordinary people could own more than one suit of clothing manufactured on large industrial machines. Earlier, most people in most parts of the world had darned or mended the one set of garments they would continue to wear through much of their lives. Odd as it sounds, this new surfeit of mass-produced clothing would deeply affect the thinking of logicians, as would the other mechanical wonders of the industrial age (Shenefelt and White 2013).

The connection was certainly explicit when Charles Babbage designed his *Analytical Engine* in analogy with the Jacard loom. Boole, did not go so far, but his work was essential to Babbage's revolutionary step. Indeed Boole framed, for the first time, logical deduction in terms of a fail-safe *general method* which would deliver the solution to all (solvable by those means) inferential questions. In a letter to Lord Kelvin dated 2 January 1851, reported by Stephen Hawking in Hawking (2006), Boole writes:

> I am now about to set seriously to work upon preparing for the press an account of my theory of Logic and Probabilities which in its present state I look upon as the most valuable if not the only valuable contribution that I have made or am likely to make to Science and the thing by which I would desire if at all to be remembered hereafter.

This view of logic appears very timely and inspiring to us, for two reasons. First, Boole tied logic to mechanical reasoning in a way which makes him the digital pioneer celebrated across science and culture. Second, and little-known outside specialist circles, he did so in a way which encompasses uncertain inference from the start. As noted above, this aspect was soon to be sacrificed to the gods of metamathematics, but it is central to the present challenges raised the across-the-board expansion of AI.

1.6 Summary of the Contributions

The three chapters by Alena Vencovská, Felix Weitkämper and Jon Williamson investigate how to assign probabilities to the sentences of a logical language. The sentences of the language are the bearers of uncertainty, hence represent the things one wants to reason about. The probabilities represent the point of view of an ideally rational uncertain reasoner. Much has been said, and surely much more is to be said, about what exactly a *rational* reasoner is or how a rational reasoner ought to behave. Rather than discussing pros and cons of various takes on the notion of rationality yet again, the authors provide expositions of well-established, logic-based approaches to uncertain reasoning.

Alena Vencovská's *Pure Inductive Logic* (Chap. 2) puts forward mathematical considerations for adopting probabilities in the absence of all domain knowledge, which can later be incorporated by some procedure to update probabilities by conditionalisation and/or Jeffrey updating. Pure Inductive Logic, is an abstract mathematical approach investigating axioms that putatively capture how a rational agent assigns probabilities to the sentences of a first-order logic. These principles are often motivated by intuitions concerning symmetries (renaming constants and relation symbols in the absence of all information about them ought not change probabilities), irrelevance (conditional probabilities ought not change in the face of irrelevant information) and relevance (the more we've seen a thing happen in the past, the more likely we think it is, that it will happen again). Surprisingly, it has been shown that principles concerning irrelevance in some sense entail principles of symmetry.

Jon Williamson's *Where do we Stand on Maximal Entropy?* (Chap. 3) reiterates a question raised by Edwin Jaynes about how to choose probabilities given a body of available evidence, but in the context of first-order logic. The Principle of Maximum Entropy holds that a rational agent ought to adopt the probability distribution with maximum entropy from all those that fit the evidence. It satisfies a number of symmetry, relevance and irrelevance axioms. Initially suggested in 1957 for finite domains, there has been recent interest in applying the principle to first-order logic, where it needs generalising to *maximal* entropy. Of particular importance are results demonstrating the decidability of a large class of inferences, in which the evidence is given by premisses and the conclusion of the inference may contain quantifiers. These facilitate fast computations of probabilities with graphical probabilistic models.

Felix Weitkämper's *Probability logic and statistical relational artificial intelligence* (Chap. 4) explores relationships between probability logic and statistical relational artificial intelligence and focuses on a number of probabilistic logics and their relationships to AI. The goal is to profit from enriching relational logic with ideas from probabilistic graphical models. Furthermore, it outlines how classical probability logic may be fruitful for the study and use of statistical relational approaches.

The two chapters by Francesco Facciuto and Sandy L. Zabell are devoted to current statistical practices with an emphasis on how we ought to reason with data. While Francesco Facciuto's chapter investigates theoretical bounds on the quality of statistical inferences, Sandy L. Zabell's chapter is interested in the quality of DNA-based statistical inferences occurring in actual legal practise.

Francesco Facciuto's *An Overview of the Generalization Problem* (Chap. 5) focusses on supervised machine learning where classifiers are presented with a data set of examples which have a label. Based on this data the classifier seeks to correctly predict labels of unseen examples. The problem which arises naturally is how to come up with a prediction method for which we have good reasons to believe that it will have a low error rate in the future. This chapter investigates how information theory can help in establishing theoretical bounds on the accuracy of such a prediction method.

Sandy Zabell's *The logic of DNA identification* (Chap. 6) deals with a topic which has become an integral part of the judicial system. Many of us will immediately think about the possibility the identity of a suspect based on a DNA sample with high probability. However, DNA evidence also has further uses such paternity testing, identifying bodies in mass disasters and historical investigations. This chapter presents the varying statistical inference procedures which are used in different applications. It thus sheds a light on the logical foundations of statistical inference.

The two chapters by Manganini & Primiero and Landes et al. both deal with the problem of biased data. Manganini & Primiero develop logical approaches to reasoning with biased and unbiased data for machine learning classifiers. Landes et al. instead show how recent work in knowledge representation and reasoning can model non-monotonic defeasible logics that under-write statistical inference procedures.

Chiara Manganini and Giuseppe Primiero's *Reasoning with and About Bias* (Chap. 7) focusses on the prediction errors which are inevitable in practical applications of machine learning. The ubiquity of machine learning algorithms raises the urgent need to minimise those errors to the largest possible extent, and otherwise to reason with them in a principled way. This chapter distinguishes between errors due to insufficient data, and sub-optimal prediction methods. It puts forward key desiderata for a logic modelling the prediction errors of machine learning classifiers and develops a framework implementing these desiderata.

Jürgen Landes, Esther Anna Corsi and Paolo Baldi's *Knowledge Representation, Scientific Argumentation and Non-monotonic Logic* (Chap. 8) starts from the observation that scientific inference is becoming more and more data-centric. While progress in artificial intelligence applications in number crunching tasks has been spectacularly successful and widely communicated, AI researchers have also made great strides developing knowledge representation and reasoning approaches. This chapter shows how formal argumentation theory developed in AI can capture non-monotonic logics, which had previously been of much interest in AI. These logics are shown to under-write statistical hypothesis testing methodology.

Finally, Maria Luisa Dalla Chiara, Roberto Giuntini and Giuseppe Sergioli's (Chap. 9) focusses on the promising approach offered by *Quantum computing* which exploits the probabilistic nature of quantum mechanics. The underlying quantum logic is thus a prime candidate for non-classical reasoning concerning uncertainty and ambiguity. This chapter shows how quantum-logical semantics can be applied for quantum machine learning. It shows that the quantum approach is particularly suitable for learning the concept of an object.

References

Ben-David, S., P. Hrubeš, S. Moran, A. Shpilka, and A. Yehudayoff. 2019. Learnability Can Be Undecidable. *Nature Machine Intelligence* 1: 44–48. https://doi.org/10.1038/s42256-018-0002-3.

Cartwright, N., J. Hardie, E. Montuschi, M. Soleiman, and A. Thresher. 2022. *The Tangle of Science*. Oxford: Oxford University Press.

Chow, S. L. 1997. *Statistical Significance*. New York: Sage.

Currie, A. 2015. Marsupial Lions and Methodological Omnivory: Function, Success and Reconstruction in Paleobiology. *Biology & Philosophy* 30 (2): 187–209. https://doi.org/10.1007/s10539-014-9470-y.

Denœux, T., D. Dubois, and H. Prade. 2020. A Guided Tour of Artificial Intelligence Research. In *Representations of Uncertainty in Artificial Intelligence: Probability and Possibility*, ed. Marquis Pierre, Papini Odile, and Prade Henri. chap. 3, 69–117. Cham: Springer. https://doi.org/10.1007/978-3-030-06164-7_3.

Feferman, A. B., and S. Feferman. 2004. *Alfred Tarski: Life and Logic*. Cambridge: Cambridge University Press.

Gabbay, D.M. 1995. *What Is a Logical System?* Oxford: Oxford University Press.

Hawking, S. 2006. *God Created The Integers: The Mathematical Breakthroughs that Changed History*. New York: Penguin.

Hansson, S.O. 2014. *David Makinson on Classical Methods for Non-Classical Problems*. Berlin: Springer.

Hey, T., K. Butler, S. Jackson, and J. Thiyagalingam. 2020. Machine Learning and Big Scientific Data. *Philosophical Transactions of the Royal Society A: Mathematical, Physical and Engineering Sciences* 378 (2166). https://doi.org/10.1098/rsta.2019.0054.

Hofmann, S.G. 2002. Fisher's Fallacy and NHST's Flawed Logic. *The American Psychologist* 57 (1): 69–70. https://psycnet.apa.org/doi/10.1037/0003-066X.57.1.69.

Krueger, J. 2001. Null Hypothesis Significance Testing: On the Survival of a Flawed Method. *American Psychologist* 56 (1):16–26. https://psycnet.apa.org/doi/10.1037/0003-066X.56.1.16.

Lewens, T. 2015. *The Meaning of Science*. London: Pelican Books.

Nowotny, H. 2021. *In AI We Trust: Power, Illusion and Control of Predictive Algorithms*. Cambridge: Polity.

Patriota, A.G. 2018. Is NHST Logically Flawed? Commentary on: "NHST Is Still Logically Flawed." *Scientometrics* 116 (3), 2189–2191. https://doi.org/10.1007/s11192-018-2817-4.

Pólya, G. 1968. *Patterns of Plausible Inference*. 2nd ed. Princeton: Princeton University Press.

Ramsundar, B., P. Eastman, P. Walters, and V. Pande. 2019. *Deep Learning for the Life Sciences: Applying Deep Learning to Genomics, Microscopy, Drug Discovery, and More*. Sebastopol: O'Reilly.

Rodriguez, S., C. Hug, P. Todorov, N. Moret, S. Boswell, K. Evans, G. Zhou, N. Johnson, B. Hyman, P. Sorger, M. Albers, and A. Sokolov. 2021. Machine Learning Identifies Candidates for Drug Repurposing in Alzheimer's Disease. *Nature Communications* 12 (1). https://doi.org/10.1038/s41467-021-21330-0.

Royall, R. 1997. *Statistical Evidence: A Likelihood Paradigm*. London: Chapman & Hall.

Senior, A.W., R. Evans, J. Jumper, J. Kirkpatrick, L. Sifre, T. Green, C. Qin, A. Žídek, A.W.R. Nelson, A. Bridgland, H. Penedones, S. Petersen, K. Simonyan, S. Crossan, P. Kohli, D.T. Jones, D. Silver, K. Kavukcuoglu, and D. Hassabis. 2020. Improved Protein Structure Prediction Using Potentials from Deep Learning. *Nature* 577 (7792): 706–710. https://doi.org/10.1038/s41586-019-1923-7.

Sim, I. 2016. Two Ways Of knowing: Big Data and Evidence-Based Medicine. *Annals of Internal Medicine* 164 (8): 562–563. https://doi.org/10.7326/M15-2970.

Shenefelt, M., and H. White. 2013. *If A, Then B: How the World Discovered Logic*. New York: Columbia University Press.

Tarski, A. 1941. *Introduction to Logic and to the Methodology of Deductive Sciences*. Oxford: Oxford University Press.

Wagner, C.G. 2004. Modus Tollens Probabilized. *British Journal for Philosophy of Science* 55: 747–753. https://doi.org/10.1093/bjps/55.4.747.

Wang, H., T. Fu, Y. Du, W. Gao, K. Huang, Z. Liu, P. Chandak, S. Liu, P. Van Katwyk, A. Deac, A. Anandkumar, K. Bergen, C.P. Gomes, S. Ho, P. Kohli, J. Lasenby, J. Leskovec, T. Liu, A. Manrai, and M. Zitnik. 2023. Scientific Discovery in the Age of Artificial Intelligence. *Nature* 620 (7972): 47–60. https://doi.org/10.1038/s41586-023-06221-2.

Chapter 2
Pure Inductive Logic

Alena Vencovská

Abstract This contribution concerns recent developments in pure inductive logic (PIL), as conceived and initially developed by Rudolf Carnap and his coworkers during the 1940s to 1970s. The basic problem of PIL is how to select an initial probability function on sentences of a given logical language so that the values it assigns to sentences, as a whole, are rational, and such that the probabilistic updating on received information yields further probability functions which are also rational in view of what the conditioning sentences are. We explain the basic notions of PIL and discuss some principles of symmetry, relevance and irrelevance as well as the principle of language invariance, and we clarify the relationship between the strongest symmetry and irrelevance principles. We illustrate the effect of various principles on the behaviour of a hypothetical virtual agent who is assumed to possess exclusively the information that is spelled out, and no other, and who aspires to be rational.

Keywords Pure inductive logic · Principles of symmetry · Invariance principle · Spectrum exchangeability · Language invariance · Principle of induction

2.1 Introduction

Pure inductive logic (PIL) is a theory of assigning probabilistic belief values to sentences in a rational way. It is a part of mathematical logic and pure mathematics where the basic framework is clearly and formally spelled out, and within this framework, additional assumptions are analysed for their consequences strictly by the means of logic and mathematics. The principles of PIL are abstracted from considerations of how people try to form reasonable beliefs on the basis of available information and from considerations of the rationality of practical decisions under

A. Vencovská (✉)
The Open University, Milton Keynes, UK
e-mail: alenavencovska1@gmail.com

H. Hosni, J. Landes (eds.), *Perspectives on Logics for Data-driven Reasoning*,
Logic, Argumentation & Reasoning 35,
https://doi.org/10.1007/978-3-031-77892-6_2

uncertainty. In this sense PIL draws on considerations of belief formation and Bayesian epistemology. However, regardless of whether or to what extent people adhere to probabilism and observe these arguably rational ideal principles in their reasoning, after the principles have been abstracted and mathematically formulated, PIL studies how—if consistently adopted—these possible principles logically determine coherent assignments of probabilistic belief values. As such, PIL is a study of logical probability. The strength of PIL lies in the understanding that we can gain from it, of the logical structure of probabilistic entailment and of connections that may be captured within this framework.

If returning with PIL to practical situations, any potential application would have to be considered on an individual basis. One assumption always present in PIL is that all the potentially relevant information is being taken account of; this is hardly ever possible in applications, so to start with a decision would have to be made about what it is that describes the situation 'sufficiently' for any given aim. Armed with that, one could try to decide which principles of PIL to adopt and pick a prior probability function satisfying them; conditioning it on some further information (evidence) should yield a probability function that inherits the rational justification of the prior and can be understood as far as logical consequences of the original principles for the conditioned function are concerned.

In what follows we will not attempt to give examples of any real possible applications of PIL since we wish to avoid discussions about the legitimacy of assuming in any particular case that the probabilistic reasoning is the right approach and that all the relevant information has been spelled out. Instead, for most of our examples we will consider a theme of a new head teacher (who would of course come to his or her school with a vast amount of relevant knowledge) and entertain the idea of this head teacher creating a virtual colleague who only gets the explicitly stated information and who argues probabilistically on the basis of certain explicitly stated principles. This is not an attempt to propose that the head teacher could be in any sense or respect imitated by such an artificial agent; we consider that impossible. It is meant merely as an illustration of how the principles which we wish to discuss can be understood in association with human reasoning, and how they work in idealised situations.

This article is largely based on the book *Pure Inductive Logic* by Jeff Paris and the author, see Paris and Vencovská (2015), where detailed proofs of most results which we mention can be found, as well as an extensive bibliography. I would like to thank Jeff Paris for helpful comments on a draft of this contribution, and also to Jonathan Weisberg and Richard Pettigrew for their helpful comments on a much earlier draft of it.

2.2 The Basic Framework

We restrict our attention to sentences of the predicate calculus coming from a language L with finitely many relation (predicate) symbols $R_1, \dots, R_q$ of arities $r_1, \dots, r_q$ respectively and countably many constant symbols $a_1, a_2, a_3, \dots$, using the usual logical connectives and quantifiers. *L does not contain equality* nor any function symbols.

Our languages thus always include the same set of constant symbols but different relation symbols and we sometimes write $L = \{R_1, \dots, R_q\}$ and refer to L as a finite language. If all the $R_1, \dots, R_q$ are unary then L is called a unary language; otherwise it is referred to as a general, polyadic language. Let SL be the set of all sentences (formulae with no free variables) of the language L and let $QFSL$ be the set of all quantifier free sentences of the language L. For simplicity (to avoid frequent double subscripts) we introduce a convention whereby b_i, b_i' can stand for one of $a_1, a_2, \dots$ so we write for example $b_1, b_2, \dots, b_m$ for $a_{i_1}, a_{i_2}, \dots, a_{i_m}$ where $i_1, \dots, i_m$ are some *distinct* natural numbers. Also, we sometimes denote relations by R, Q, P.

By a structure for L with universe (domain) $\{a_1, a_2, a_3, \dots\}$ we mean the usual model-theoretic notion: the set $\{a_1, a_2, a_3, \dots\}$ with each a_i being the interpretation of itself along with a set of r_j-tuples from $\{a_1, a_2, a_3, \dots\}^{r_j}$ for each relation symbol R_j with arity r_j, interpreting R_j.

The reason for choosing this setup is that it corresponds to many situations about which one may wish to argue: it suffices to describe what is important for a rational agent who interprets what they encounter in terms of (an unlimited number of) individuals and (a fixed number of) their properties or relations between them.

Example 2.1 Consider a new head teacher in a secondary school and her reasoning about the students of the school, their properties and relations between them. Clearly, her language is not like the one above. Students have names like

> Martin Brown, Gemma Derwent, John Wendig, ...

and their properties and the relations between them are expressed in natural language as

> being good at PE (about 1 student),
> making a good relay race team (relating 4 students)

and so on. Even before she gets any information about students, relay race teams and other matters at the school, on which she will build her opinions about school affairs, the words tell her a lot: 'Gemma' unlike 'Martin' and 'John' is a female name, having somebody good at PE will make a quadruple of students more likely to make a good relay race team etc.

(continued)

Example 2.1 (continued)
If she wished to compare her reasoning to something accountable, she—perhaps being an IT enthusiast—might set up a virtual agent that does use the language as above, and consider how it behaves. The agent would have to make conclusions merely on the basis of various (sets of) sentences, without 'knowing' of any intended interpretation of constants and relation symbols used. Thus although the head teacher might intend $a_1, a_2, a_3, \ldots$ to stand for *Martin, Gemma, John, . . .* respectively and

$R_1(x), R_2(x, y, z, u), \ldots$

for

x is good at PE,
x, y, z, u would make a good relay race team

etc. respectively, she would give the agent information about the students expressed in the agent's formal language and the agent could not use any information that has not been spelled out. The ambition would be to set up the agent so that in particular situations, if enough relevant information was collected and expressed in the agent's language, the conclusions suggested by him would be reasonable. The accountability of the agent would be based on the fact that he got parameters for his initial probability distribution, chosen to satisfy various principles of PIL, from the head teacher, and she would be able to check how it changes as he gets further information (always expressed in his artificial language so contributing just what it states and nothing else). The way in which the agent's probability distribution would change would depend on the information received in a way governed by the principles satisfied by the agent's original probability distribution.[1] For the head teacher, the agent's developing probability distribution would translate back to refer to her school and could be judged as reasonable or not.

We will use this idea to illustrate how various conditions imposed on the virtual agent influence his conclusions. We will call the agent Phil. For simplicity and contrast we refer to the head teacher as feminine and to the agent Phil as masculine.

2.2.1 Probability Functions

We are interested in assigning rational, or logical, belief values to sentences of a language as above. A value needs to be assigned to each sentence so that the assignment as a whole is reasonable. This yields some intuitive rationality

[1] Even though some principles are not preserved after conditioning.

requirements e.g. of the type *'if A gets such and such a value then B should get the same, larger, smaller or whatever value'* in the light of how A and B are related. Some particular rationality requirements lead to arguments in favour of logical belief values being probabilistic, see for example Hajek and Hall (2002), Joyce (2009), Pettigrew (2011), Savage (1954), and Paris and Vencovská (2015). That is, in favour of rational belief assignments being probability functions defined on the set SL where:

Definition 2.1 A function $p : SL \to [0, 1]$ is a probability function on SL if for all A, B and $\exists x\, D(x) \in SL$

(P1) If A is logically valid[2] [$\models A$] then $p(A) = 1$.
(P2) If A and B are mutually exclusive [$\models \neg(A \wedge B)$] then

$$p(A \vee B) = p(A) + p(B)\,.$$

(P3) $p(\exists x\, D(x)) = \lim_{n\to\infty} p(D(a_1) \vee D(a_2) \vee \ldots \vee D(a_n))$.

Thus when talking about an assignment of belief values to sentences of L we will mean choosing a probability function on SL and we will use the terms belief function and probability function synonymously throughout.

We note that any function $p : QFSL \to [0, 1]$ satisfying just (P1) and (P2) has a unique extension[3] to a probability function on SL so in many situations it suffices to think of probability functions as defined on $QFSL$ and satisfying (P1) and (P2). In particular to specify a probability function it suffices to give its values for quantifier free sentences and as we shall shortly see, this can be further reduced to a special class of such sentences called state descriptions.

(P1) and (P2) ensure that p has 'common sense' properties; in particular for $A, B \in SL$ we have

- $p(\neg A) = 1 - p(A)$.
- If A is contradictory then $p(A) = 0$.
- If A logically implies B then $p(A) \leq p(B)$.
- If A is logically equivalent to B then $p(A) = p(B)$.

Henceforth we will use these properties without further mention. In particular we take it as a matter of course that the same belief values are assigned to logically equivalent sentences, in effect arguing about sentences *up to logical equivalence*.

---

[2] Equivalently, if A is provable, or again equivalently, if A is true in every structure for L with universe $\{a_1, a_2, \ldots\}$ with each constant symbol a_i interpreted as itself, see Paris and Vencovská (2015). A similar remark applies to (P2).

[3] By Gaifman's Theorem, see Gaifman (1964).

2.2.2 *Conditional Probability*

In situations that motivate pure inductive logic, rational agents mostly need to form belief values on the basis of some evidence. PIL adopts the stance, rationally justified by a conditional Dutch Book argument, see for example Paris and Vencovská (2015, Chapter 5), that belief values formed on the basis of some evidence should be conditional probabilities.

Conditional probability is defined as follows.

Given a probability function p on SL and $B \in SL$ with $p(B) > 0$, the *conditional probability function* $p(\,.\,|B) : SL \to [0, 1]$ is defined by

$$p(A|B) = \frac{p(A \wedge B)}{p(B)}. \tag{2.1}$$

The conditional probability gives belief values that again conform to the requirements (P1)–(P3): if p is a probability function on SL, $B \in SL$ and $p(B) > 0$ then $p(\,.\,|B)$ is a probability function and $p(A|B) = 1$ whenever $B \models A$.

In order to avoid some formal problems with 0 we embrace the frequently adopted convention that expressions such as $p(D \mid E) = c$ mean $p(D \wedge E) = c\, p(E)$ etc. regardless of the value of $p(E)$.

2.2.3 *The Basic Question*

Upon accepting that rational belief values should be probabilistic and updating them on the basis of some information should be done via conditional probability, the fundamental question can be stated as follows: *which probability functions represent rational systems of belief values, also when conditional probabilities are taken to represent the belief values on the basis of some evidence?*

Thus the idea is that with a given language L, a rational agent should be able to choose a probability belief function $p : SL \to [0, 1]$ that represents their beliefs on the basis of no information, or evidence, but taking into account what it would mean should some information arise, since then they would be committed to the belief values obtained from the original probability function by conditioning on this evidence.

Example 2.2 Returning to our new head teacher, according to the above arguments, Phil would simply be a probability distribution p on a sufficiently rich language L.

(continued)

Example 2.2 (continued)
If our head teacher wished to consider sentences involving just one unary property in isolation, for example

x prefers Physics to Chemistry, (*)

then she would need L to contain one unary predicate $R(x)$. She would express her information about students as $A(b_1, \ldots, b_m)$, with $b_1, \ldots, b_n$ replacing the names of the distinct students involved and $R(x)$ replacing (*). Similarly she would express any other statement to which she wishes to attach a belief value, obtaining some $B(b'_1, \ldots, b'_k)$; Phil's advice then would be to give this statement the belief value

$$p\left(B(b'_1, \ldots, b'_k) \mid A(b_1, \ldots, b_m)\right)$$

(provided $p(A(b_1, \ldots, b_m)) \neq 0$).

An initial reaction to the question what probability function should a rational agent hold when nothing is known might be that any instantiation of a relation $Q(b_1, \ldots, b_s)$ or its negation $\neg Q(b_1, \ldots, b_s)$ should get belief value $\frac{1}{2}$, and that they should be mutually stochastically independent, so any conjunction of k such distinct[4] instantiations or their negations should get value $\frac{1}{2^k}$. The problem with the probability function which is thereby determined is that no matter what information we conditionalise it with, the instantiations of relations which are not decided by the information will still get value $\frac{1}{2}$ in spite of potentially vast amounts of evidence pointing the same way. So for example an agent working with a unary R continues to have the belief value $\frac{1}{2}$ for $R(b_n)$ even on the basis of $R(b_1) \wedge \ldots \wedge R(b_{n-1})$ no matter how large n is.

The apparently obvious answer is thus not helpful. It seems irrational to commit oneself to ignoring a vast amount of evidence pointing the same way. Yet considering where the answer came from, we see that it arose from two arguably rational requirements:

- *symmetry* between instantiations of relations and their negations—note that p being a probability function means that requiring

$$p(R(b_1, \ldots, b_s)) = \frac{1}{2}$$

is the same as requiring

$$p(R(b_1, \ldots, b_s)) = p(\neg R(b_1, \ldots, b_s))$$

[4] 'Distinct' meaning that they vary either in the relation or some constant.

- *irrelevance* of instantiations of a relation or its negation to other ones—requiring

$$p(R(b_1,\ldots,b_s)\wedge Q(b'_1,\ldots,b'_t)) = p(R(b_1,\ldots,b_s))\cdot p(Q(b'_1,\ldots,b'_t))$$

  amounts to $p(Q(b'_1,\ldots,b'_t)) = 0$ or

$$p(R(b_1,\ldots,b_s)\mid Q(b'_1,\ldots,b'_t)) = p(R(b_1,\ldots,b_s)),$$

  so unless $Q(b'_1,\ldots,b'_t)$ is judged impossible it is irrelevant to $R(b_1,\ldots,b_s)$. Similarly for conjunctions of $R(b_1,\ldots,b_s)$, $Q(b'_1,\ldots,b'_t))$ or their negations.

It is the latter which in particular appears too strong and more contentious. The natural next step thus is to consider other rational requirements, or *principles*, that can be imposed on p, and see if some selection of them determines a probability function that is a good starting point for reasoning in the above sense. This is where PIL really starts, formulating and investigating rational principles for belief assignment that should augment the basic requirement that these values should constitute a probability function. To date, no clear winner has emerged out of these considerations. In the special case of unary inductive logic, that is, when all the predicates are unary, there are the long advocated Carnap Continuum functions based mainly on Johnson's Sufficientness Principle, see Johnson (1932), but another unary continuum of probability functions justified by other principles does exist, see Paris and Vencovská (2015, Chapter 18,19), and the general case of languages with relations of arbitrary arities is even more intriguing. It may yet turn out that the general case explains much about the unary case, in the way complex numbers do about the reals.

2.3 Principles of Pure Inductive Logic

Principles of PIL can be grouped according to their motivation; it might stem from considerations of symmetry, relevance, irrelevance, analogy or from some other source. Such groupings are mainly for the sake of exposition since it is quite often possible to see a certain principle as motivated by more than one of these types of considerations so the classification is somewhat arbitrary. Furthermore, motivation for principles happens on an intuitive rather than formal level.[5]

In this part we shall summarise some of the principles that have been proposed. First however we shall introduce the concept of a state description necessary for the formulation of some of the principles.

[5] We remark that in the case of symmetry, it is possible in PIL to formalise the intuition quite well, so that we have a precise formal criterion for when a principle should count as a symmetry principle. However, it is somewhat technical. We state it at the end of Sect. 2.3.2.

2.3.1 State Descriptions

A state description for $b_1, \ldots, b_m$ in a language $L = \{R_1, \ldots, R_q\}$, with $R_1, \ldots, R_q$ of arities $r_1, \ldots, r_q$ respectively, is any sentence of the form

$$\bigwedge_{i=1}^{q} \bigwedge_{\langle j_1,\ldots,j_{r_i}\rangle \in \{1,\ldots,m\}^{r_i}} \pm R_i(b_{j_1}, b_{j_2}, \ldots, b_{j_{r_i}}) \tag{2.2}$$

where $\pm R_i(b_{j_1}, b_{j_2}, \ldots, b_{j_{r_i}})$ stands for one of

$$R_i(b_{j_1}, b_{j_2}, \ldots, b_{j_{r_i}}), \quad \neg R_i(b_{j_1}, b_{j_2}, \ldots, b_{j_{r_i}}).$$

State descriptions will usually be denoted by capital Greek letters Θ, Φ, Ψ etc.

Hence distinct state descriptions for $b_1, \ldots, b_m$ are mutually exclusive sentences each of which fully describes which relations do or do not hold as regards the $b_1, \ldots, b_m$. Replacing the constants in a state descriptions by variables yields a *state formula*.

Example 2.3 If L contained just two relation symbols, a unary $R(x)$ and a binary $Q(x, y)$ then one state description for a_1, a_3 in this language would be

$$R(a_1) \wedge \neg R(a_3) \wedge Q(a_1, a_1) \wedge Q(a_1, a_3) \wedge Q(a_3, a_1) \wedge \neg Q(a_3, a_3)\,.$$

The other $2^6 - 1$ state descriptions for a_1, a_3 in this language are obtained by placing negations to the above two instantiations of R and four instantiations of Q differently. If a head teacher wished to interpret $R(x)$ as

'x votes for introducing a new school uniform',

and $Q(x, y)$ as

'x likes y in the new school uniform'

and a_1, a_3 as Martin and John respectively, then the above state description would express that Martin votes for the new school uniform but John does not, Martin likes both himself and John in the new school uniform and John likes Martin in it but not himself.

In a language with only *unary* relation symbols $R_1, \ldots, R_q$ state descriptions for $b_1, \ldots, b_m$ have a particularly simple form:

$$\bigwedge_{j=1}^{m} \bigwedge_{i=1}^{q} \pm R_i(b_j)$$

The formulae $\bigwedge_{i=1}^{q} \pm R_i(x)$ are called atoms; they are usually denoted $\alpha_1(x), \alpha_2(x) \ldots, \alpha_{2^q}(x)$, listed lexicographically from all positives to all negatives. Unary state descriptions are usually written as

$$\bigwedge_{j=1}^{m} \alpha_{k_j}(b_j) \tag{2.3}$$

More generally, in a polyadic language, (polyadic) atoms are state formulae for r variables, where r is the highest arity of a relation symbol in the language. As in the unary case, any state description can be expressed as a conjunction using just atoms instantiated by r-tuples of constants involved, but unlike the unary case not all such conjunctions are state descriptions because only some are consistent.

An important property of probability functions is that to specify them we only need to specify their values for state descriptions. This is because by the Disjunctive Normal Form Theorem every quantifier free sentence involving constants $b_1, \ldots, b_m$ is logically equivalent to a disjunction of state descriptions for these constants in the given language, and since the state descriptions for $b_1, \ldots, b_m$ are clearly pairwise mutually exclusive, the probability value of the sentence must be the sum of the values of the state descriptions featuring in its disjunctive normal form representation. And as we have already mentioned, the probability values given to quantifier free sentences uniquely determine the probability function for all sentences.

2.3.2 Symmetry Principles

Constant Exchangeability, Ex *Let $A(b_1, \ldots, b_m) \in SL$ and let $b_1', \ldots, b_m'$ be any other choice of distinct constant symbols from amongst the $a_1, a_2, \ldots$ Then*

$$p(A(b_1, \ldots, b_m)) = p(A(b_1', \ldots, b_m'))\,. \tag{2.4}$$

Predicate Exchangeability, Px *If R, Q are relation symbols of L with the same arity, $A \in SL$ and A' the result of simultaneously replacing R by Q and Q by R throughout A then*

$$p(A) = p(A')\,. \tag{2.5}$$

These principles clearly are essential for a probability function that should be adopted by a rational agent on the basis of no evidence since nothing is supposed to be known about the individuals or the relations (beyond their arities). Note that these principles are not necessarily preserved after conditioning since evidence may distinguish some individuals and some relations.

Example 2.4 Consider our head teacher and the virtual agent again. Phil's initial probability function should satisfy Ex and Px since otherwise his advice would depend on which individual symbols are chosen to represent which students, and similarly for relation symbols. In Phil's initial state of ignorance there is no rational reason to make any such distinction.

Strong Negation Principle, SN *If $A \in SL$ and A' is the result of replacing each occurrence of R in A by $\neg R$ then*

$$p(A) = p(A')\,.$$

SN is more contentious than Ex or Px. Requiring that a probability function which should be adopted by a rational agent on the basis of no evidence satisfies SN amounts to assuming, to start with, that the properties and their negations are in some sense of equal status.

Variable Exchangeability, Vx *Let R be a relation symbol of L with arity s and let σ be a permutation of the set $\{1, \ldots, s\}$. If $A \in SL$ and A' is the result of replacing each $R(t_1, t_2, \ldots, t_s)$ appearing in A by $R(t_{\sigma(1)}, t_{\sigma(2)}, \ldots, t_{\sigma(s)})$ (where $t_1, \ldots, t_s$ are constants or variables) then*

$$p(A) = p(A')\,.$$

We remark that having $A \in QFSL$ in place of $A \in SL$ in the statements of Ex, Px, SN and Vx yields equivalent principles.

Also, these principles are independent in the sense that none of them follows from the others. In particular, Vx *does not* follow from Ex.

Example 2.5 If $L = \{R, Q\}$ where R is unary and Q is binary, Vx mandates

$$p(R(a_1) \wedge \neg R(a_3) \wedge Q(a_1, a_3)) = p(R(a_1) \wedge \neg R(a_3) \wedge Q(a_3, a_1)) \tag{2.6}$$

but Ex does not.

Hence if Phil's probability distribution p on L as above satisfies Vx then his advice to a head teacher on the basis of (only) the information that Martin votes for the new school uniform but John does not, is to give an equal belief value to Martin liking John in the new uniform as John liking Martin in it.

(continued)

Example 2.5 (continued)
This is since using R, Q, a_1 and a_3 as in Example 2.3, by (2.1) and (2.6),

$$p(Q(a_1, a_3) \mid R(a_1) \wedge \neg R(a_3)) = p(Q(a_3, a_1) \mid R(a_1) \wedge \neg R(a_3))\,.$$

We take this opportunity to remark that the advice being somewhat contentious does not necessarily discredit Vx, but it highlights the problem of not spelling out enough information (since there is an implicit connection between voting for a new uniform and liking people in it).

For the case of unary languages we have

(Unary) Atom Exchangeability *For any permutation* τ *of* $\{\alpha_1, \alpha_2, \ldots, \alpha_{2^q}\}$ *and constants* $b_1, b_2, \ldots, b_m$,

$$p\left(\bigwedge_{j=1}^{m} \alpha_{k_j}(b_j)\right) = p\left(\bigwedge_{j=1}^{m} \tau(\alpha_{k_j})(b_j)\right). \tag{2.7}$$

Example 2.6 Suppose that our head teacher consults Phil about sentences involving merely the unary properties
'x gets B or better in Maths on the end-of-year report',
and
'x gets B or better in English on the end-of-year report'.

They would split a class of students into four groups, and if Phil's probability function p on SL for an appropriate language satisfied Ax and Ex then Phil would advise to give the same probability to outcomes where these groups can be matched merely for size (irrespective of the performance of the groups which are being matched).

For general languages, we forgo the precise formulation of the corresponding principle, stating it merely as

Permutation Invariance Principle, PIP *For any* admissible *permutation of (polyadic) atoms, the probability of a state description equals the probability of the state description obtained from the original one by replacing each atom by its image.*

See Paris and Vencovská (2015), Paris and Vencovská (2019), and Ronel and Vencovská (2014) for details and for equivalent formulations of this principle.

All the above principles can be phrased as requiring that the probability values are invariant under certain permutations of sentences respecting logical equivalence.

For example for SN they are the permutations η_R defined by

$$\eta_R(A) = A' \quad \text{for } A \in SL$$

where A' is the result of replacing each occurrence of a relation symbol R in A by $\neg R$, and Ex amounts to the requirement that p be invariant under all η_σ where σ is a permutation of $\{1, 2, 3, \ldots\}$ and

$$\eta_\sigma(A(a_1, \ldots, a_m)) = A(a_{\sigma(1)}, \ldots, a_{\sigma(m)})) \quad \text{for } A(a_1, \ldots, a_m)) \in SL\,.$$

(Clearly, these permutations do respect logical equivalence.)

PIP can also be phrased as a principle requiring that probability values are invariant under certain permutations of sentences, including all those involved in the case of SN, Px and Vx, and as such it is stronger than these principles. The combination PIP+Ex might appear to be the strongest symmetry requirement we should make. However, there are still other permutations of sentences that arguably should be respected, namely all those that are *grounded in permutations of structures* for the given language in the following sense.

Let $\mathcal{T}L$ be the set of structures for L with domain $\{a_1, a_2, \ldots\}$, and each constant symbol a_n interpreted as itself. A permutation η of SL is grounded if it respects logical equivalence and if there is a permutation f of $\mathcal{T}L$ (a bijection of $\mathcal{T}L$ onto itself) such that

$$\{f(M) : M \models A\} \;=\; \{M \in \mathcal{T}L : M \models \eta(A)\} \;\; \text{for all } \; A \in SL\,. \tag{2.8}$$

For imagine that the rational agent's task to assign belief values to sentences arises through them finding themselves in one of the structures for L with the universe $\{a_1, a_2, \ldots\}$, knowing that this is so and aware of the possibilities (of $\mathcal{T}L$ and its subsets $\{M : M \models A\}$ for $A \in SL$), but knowing nothing beyond this; in particular, ignorant of which structure they are in. Then we can argue that whatever probability belief function they should settle on it should be invariant under all permutations of sentences grounded in the above sense. The reason is that from a specific M, the agent would see $\mathcal{T}L$ with its subsets defined by various A in the same way as they would see $\mathcal{T}L$ with the corresponding subsets defined by $\eta(A)$ from $f(M)$.

These considerations lead to a principle which strengthens PIP+Ex:

Invariance Principle, INV *Assume that η is a permutation of sentences grounded in the above sense. Then*

$$p(A) = p(\eta(A)) \quad \text{for all } A \in SL.$$

However, this principle is just too strong, see Paris and Vencovská (2011) and Paris and Vencovská (to appear), in the sense that it is satisfied by just one probability function on SL, namely by Carnap's c_0 in the unary case and by its obvious generalisation for polyadic languages (where the only state descriptions with nonzero probability – the same for any of them – are those with all constants featuring in it satisfying the same relations).

Any symmetry principle we have mentioned above can be defined in terms of respecting an appropriate subset of the set of all permutations of sentences grounded in permutations of structures. There are yet other subsets of the set of all permutations of sentences grounded in permutations of structures that lead to new principles. Within PIL, we consider the existence of a set of permutations of sentences grounded in permutations of structures such that a principle can be expressed in terms of respecting permutations from this set to be the precise definition of what it means to be a symmetry principle.

Current research indicates that the strongest reasonable symmetry principle is the combination Ex+ENV, where ENV mandates respecting permutations characterised by a certain condition which implies that the image of a sentence mentions at most those constants that are mentioned in the original sentence. Ex+ENV is strictly stronger than Ex+PIP, see e.g. Paris and Vencovská (2022) for details.

2.3.3 Irrelevance Principles

The next group of principles that come to mind when considering rational belief assignments are principles of irrelevance, or independence, in the first place

Constant Irrelevance Principle, IP *If* $A, B \in QFSL$ *have no constant symbols in common then*

$$p(A \wedge B) = p(A) \cdot p(B)$$

(Equivalently, $p(A \mid B) = p(A)$.)

IP thus says that if A mentions no constant that also appear in B then belief in A should not change when conditioning on B. This might be argued for on the grounds that being about different individuals, B tells us nothing about A.

Example 2.7 Let $L = \{R, Q\}$ again, with R unary and Q binary. Then for p satisfying IP we must have

$$p(R(a_2)) = p(R(a_2) \mid Q(a_1, a_3))$$

and

$$p(R(a_2)) = p(R(a_2) \mid R(a_1))$$

but possibly

$$p(R(a_1)) \neq p(R(a_1) \mid Q(a_1, a_3)) .$$

Interpreting a_1, a_2, a_3 as Martin, Gemma or John respectively, $Q(x, y)$ as
'x sometimes joins y for lunch'
and $R(x)$ as
'x has discipline issues'
shows that under IP, Phil's advice to the head teacher on the basis of (only) the information that Martin sometimes joins John for lunch, would be not to change the belief value for Gemma having discipline issues, although we cannot tell from IP alone what his advice would be as regards Martin having discipline issues. Phil would even advise that the belief value in Gemma having discipline issues should not change on the basis of the information that Martin has discipline issues.

The Constant Irrelevance Principle IP has been very important in *unary* inductive logic (where all predicates are unary) since the probability functions satisfying IP and Ex are particularly simple, and any probability function satisfying Ex can in a sense be expressed as a mixture of them.[6]

In polyadic inductive logic (where there are relations of higher arities) it is still the case that any probability function satisfying Ex can be expressed as a mixture[7] of probability functions satisfying IP and Ex. However, it is the following pair of irrelevance principles which has so far played a more significant role in the development of polyadic inductive logic.

[6] For a more precise statement see de Finetti's theorem, cf. de Finetti (1974) or Paris and Vencovská (2015, Chapter 9).

[7] See e.g. Paris and Vencovská (2015, Chapter 25).

Weak Irrelevance Principle, WIP *Suppose that $A, B \in QFSL$ are such that they have no constant nor relation symbols in common. Then*

$$p(A \wedge B) = p(A) \cdot p(B).$$

Example 2.8 Continuing Example 2.7, for p satisfying WIP we still must have

$$p(R(a_2)) = p(R(a_2) \mid Q(a_1, a_3))$$

but possibly

$$p(R(a_2)) \neq p(R(a_2) \mid R(a_1))$$

and (as with IP) possibly

$$p(R(a_1)) \neq p(R(a_1) \mid Q(a_1, a_3))\,.$$

Hence WIP unlike IP does not oblige Phil to ignore evidence concerning Martin having discipline issues when it comes to the belief value for Gemma having discipline issues.

To state the second principle mentioned above we need some more notation. Working again with our fixed general language L, let

$$\Theta(b_1, \ldots, b_m) = \bigwedge_{i=1}^{q} \quad \bigwedge_{\langle j_1, \ldots, j_{r_i} \rangle \in \{1, \ldots, m\}^{r_i}} \pm R_i(b_{j_1}, b_{j_2}, \ldots, b_{j_{r_i}}) \tag{2.9}$$

be a state description. Let $s, t \in \{1, \ldots, m\}$. We say that b_s and b_t are *indistinguishable* with respect to Θ and write $b_s \sim_\Theta b_t$ if there is nothing in the state description to tell them apart. Precisely, $b_s \sim_\Theta b_t$ holds if for any $i \in \{1, \ldots q\}$ and not necessarily distinct $j_1, \ldots, j_u, j_{u+2}, \ldots, j_{r_i}$ from $\{1, 2, \ldots, m\}$ (possibly including some further occurrences of s, t),

$R_i(b_{j_1}, \ldots, b_{j_u}, b_s, b_{j_{u+2}}, \ldots, b_{j_{r_i}})$ occurs positively in (2.9)

just when

$R_i(b_{j_1}, \ldots, b_{j_u}, b_t, b_{j_{u+2}}, \ldots, b_{j_{r_i}})$ occurs positively in (2.9).

Example 2.9 For simplicity, let our language consist of just one binary relation symbol $R(x, y)$. Let $\Theta(b_1, b_2)$ be the state description

$$R(b_1, b_1) \wedge \neg R(b_1, b_2) \wedge R(b_2, b_1) \wedge R(b_2, b_2)\,.$$

Then clearly, b_1 and b_2 are *distinguishable* with respect to Θ. They are also distinguishable, for example, with respect to

$$\neg R(b_1, b_1) \wedge R(b_1, b_2) \wedge R(b_2, b_1) \wedge \neg R(b_2, b_2)\,.$$

In fact, the only state descriptions for b_1 and b_2 for this language where they are *indistinguishable*, are

$$R(b_1, b_1) \wedge R(b_1, b_2) \wedge R(b_2, b_1) \wedge R(b_2, b_2)$$

and

$$\neg R(b_1, b_1) \wedge \neg R(b_1, b_2) \wedge \neg R(b_2, b_1) \wedge \neg R(b_2, b_2)\,.$$

We often picture the state descriptions for this language as matrices with 0,1 entries where there is 1 or 0 in the ith row, jth column if the state description contains $R(b_i, b_j)$ or $\neg R(b_i, b_j)$ respectively, so the above state descriptions correspond, in order, to

$$\begin{matrix} 1 & 0 \\ 1 & 1 \end{matrix} \qquad \begin{matrix} 0 & 1 \\ 1 & 0 \end{matrix} \qquad \begin{matrix} 1 & 1 \\ 1 & 1 \end{matrix} \qquad \begin{matrix} 0 & 0 \\ 0 & 0 \end{matrix}$$

Example 2.10 Let our language again consist of just one binary relation symbol $R(x, y)$. In the state description $\Psi(b_1, \ldots, b_5)$ corresponding to

$$\begin{matrix} 1 & 0 & 0 & 0 & 1 \\ 0 & 1 & 1 & 0 & 0 \\ 0 & 1 & 1 & 0 & 0 \\ 0 & 0 & 0 & 1 & 0 \\ 1 & 0 & 0 & 0 & 1 \end{matrix} \tag{2.10}$$

we have $b_1 \sim_\Psi b_5$ and $b_2 \sim_\Psi b_3$ but all the other pairs are distinguishable.

(continued)

Example 2.10 (continued)
We can consider an interpretation where the individuals are building blocks of children's construction sets, like Lego, Duplo and its imitations, with $R(x, y)$ standing for

'x has the right size studs to fit the holes of y',

that is (allowing for the same brick to 'fit' under itself, or not) 'x fits under y', or 'y fits on x'. A state description for m blocks then specifies exactly which block fits on which block and under which block, and indistinguishability with respect to it corresponds to fitting on and under the same blocks.

The above state description would be about five blocks from three incompatible (both ways) sets.

It is easy to see that $\sim_\Theta$ as defined above is an equivalence relation on $\{b_1, \ldots, b_m\}$. We define the *spectrum of* Θ, denoted $\mathcal{S}(\Theta)$, to be the multiset[8] of sizes of the corresponding equivalence classes.

Example 2.11 For the state description $\Psi(b_1, \ldots, b_5)$ corresponding to (2.10) the equivalence classes are $\{b_1, b_5\}$, $\{b_2, b_3\}$ and $\{b_4\}$ and the spectrum $\mathcal{S}(\Psi)$ is $\{2, 2, 1\}$.

Spectrum Exchangeability, Sx *The probability of a state description depends only on its spectrum.*

We remark that when the language is unary, Sx is equivalent to Ax+Ex.

Example 2.12 If our head teacher consults Phil about sentences involving just one binary relation

'x likes y in the new school uniform'

then provided Phil's probability function p satisfies Sx, his advice on the complete specifications of how 4 students like each other in the new uniform corresponding to matrices

$$\begin{matrix} 0 & 0 & 0 & 0 \\ 0 & 0 & 0 & 0 \\ 0 & 0 & 0 & 0 \\ 1 & 1 & 1 & 1 \end{matrix} \qquad \begin{matrix} 0 & 1 & 1 & 1 \\ 1 & 1 & 1 & 1 \\ 1 & 1 & 1 & 1 \\ 1 & 1 & 1 & 1 \end{matrix}$$

(continued)

[8] A multiset is as a set but allowing multiple instances for each of its elements.

Example 2.12 (continued)
is to give them the same belief value since the spectrum is {3, 1} in both cases. (Note that in the first case there are three students who do not like anybody in the new uniform, including themselves and one student who likes everybody in it including him/herself and in the second case there are three students who do like everybody in the new uniform, including themselves and one student who also likes everybody in it excluding him/herself.)

Example 2.13 Continuing Examples 2.10 and 2.11, the state description $\Phi(b_1, \ldots, b_5)$ corresponding to

$$\begin{matrix} 1 & 1 & 1 & 1 & 1 \\ 1 & 1 & 1 & 1 & 1 \\ 0 & 0 & 1 & 1 & 0 \\ 0 & 0 & 1 & 1 & 0 \\ 1 & 1 & 0 & 0 & 1 \end{matrix}$$

yields $b_1 \sim_\Phi b_2$ and $b_3 \sim_\Phi b_4$ with all the other pairs distinguishable, so $\mathcal{S}(\Phi) = \{2, 2, 1\}$ again and hence any probability function p satisfying Sx would have to give these state descriptions an equal value, $p(\Psi) = p(\Phi)$.

It is not so immediately clear how this state description would arise for a quality mixture of building blocks, but with a mixture of not so good imitations of Lego it could happen. The toy box owner's probability functions satisfying Sx would mean that s/he considers it equally probable that upon fishing out five blocks splitting into two pairs that behave the same and an odd block, any other fits between the five might work, or not.

Spectrum Exchangeability is one of the earliest, and most investigated principles in polyadic inductive logic. Initially, it was seen as a symmetry principle (a polyadic generalisation of Atom Exchangeability) and some effort has been made to understand how it might arise as a weakening of INV through limiting the set of permutations η to some suitable subset. Current research however suggests that this is not possible, and Sx has come to be understood as a principle of irrelevance - nothing but the spectrum is relevant for the probability of a state description. In Paris and Vencovská (to appear), it is shown that Sx is strictly stronger than Ex+ENV. As such, Sx is the strongest currently known rational principle in PIL.

Sx in combination with the principle that we are going to introduce next, Language Invariance, guarantees some particularly desirable properties for the probability functions satisfying them.

2.3.4 Language Invariance

Language Invariance, Li *A probability function p for a language L satisfies Language Invariance if there is a family of probability functions* $p^{\mathcal{L}}$, *one on each finite language* $\mathcal{L}$, *satisfying Px and Ex, such that* $p^{L} = p$ *and whenever* $\mathcal{L} \subseteq \mathcal{L}'$, $p^{\mathcal{L}}$ *and* $p^{\mathcal{L}'}$ *agree on sentences of* $\mathcal{L}$.

We say that p satisfies Language Invariance with $\mathcal{P}$, *where* $\mathcal{P}$ *is some property, if the members* $p^{\mathcal{L}}$ *of this family all satisfy also the property* $\mathcal{P}$.

This is the only principle of PIL which links different languages - all the others apply to a probability function on SL for a given fixed language. Li has proved to be surprisingly consequential and it has a strong claim to being considered a desirable rational principle to impose on a probability belief function that an agent should accept on the basis of no knowledge. Note that this principle incorporates Ex and Px. Conditioning on some sentence with non-zero probability would give the agent another coherent family of probability functions (where Px and Ex no longer need to hold), one for each finite language containing all the relation symbols that appeared in the conditioning sentence, which would all reflect the added knowledge in the same way.

We could avoid the need for Language Invariance and the perhaps strange asymmetry in allowing infinitely many constant symbols and only finitely many relation symbols by working instead with infinitely many relation symbols of each arity to start with. A language invariant family in the sense of our definition simply corresponds to one probability function on sentences of such a language and it could be recovered by restricting this probability function to the sets of sentences of the individual finite languages. However, for historical reasons, and since it seems simpler, we use the present setup.

Example 2.14 In most situations where a rational agent needs to hold a belief function on sentences of a finite language it is desirable that if necessary this could be extended to allow them to consider sentences involving other relation symbols than the original finitely many relation symbols, so that this extension process does not involve revising old belief values.

In particular, our head teacher who wished to argue about the discipline issues and joining people for lunch in Examples 2.7 and 2.8 may well find that she needs to think also of

'x helps y with Maths homework'

and Phil should have the possibility of adding a binary relation symbol and extending his probability function to be the same for sentences involving only R and Q while having a value for any sentence involving also the added symbol.

2.3.5 Principle of Induction

There are many other interesting principles, notably those of analogy and relevance, see for example Paris and Vencovská (2016) and Paris and Vencovská (2015), where also numerous relevant references can be found, but discussing them would make this contribution too long. We will now merely mention one fundamental principle of relevance.

As noted before in connection with the 'obvious' answer to the question of what probability function should a rational agent hold on the basis of no information, a rational probability function should allow for learning from experience. One interpretation of what this means is that the conditional probability of an individual having some properties should reflect what the evidence tells us about these same properties holding for other individuals; the more individuals are identified as such by the evidence, the greater the conditional probability for the new individual having these properties.

Assuming that the evidence comes in the form of a state description which, as we have seen above when introducing the principle Sx, splits the individuals concerned into indistinguishability classes that group together individuals with exactly the same properties, the above maxim finds expression in a principle stating that a new individual should be more likely to join a larger indistinguishability class than a smaller one:

Principle of Induction, PI *Let $\Theta(b_1, \ldots, b_m)$ be a state description and let $\Theta_1(b_1, \ldots, b_m, b_{m+1})$ and $\Theta_2(b_1, \ldots, b_m, b_{m+1})$ be state descriptions extending Θ. Assume that b_{m+1} is indistinguishable from some of the $b_1, \ldots, b_m$ with respect to Θ_1, and that there are fewer or at most as many of the $b_1, \ldots, b_m$ indistinguishable from b_{m+1} with respect to Θ_2. Then*

$$p(\Theta_1 \mid \Theta) \geq p(\Theta_2 \mid \Theta) \tag{2.11}$$

(equivalently, $p(\Theta_1) \geq p(\Theta_2)$).

Example 2.15 Let L be a language with only one binary relation symbol and p a probability function on SL satisfying PI. Assume that a state description $\Psi(b_1, \ldots, b_5)$ is represented by the matrix

$$\begin{matrix} 1 & 0 & 0 & 0 & 1 \\ 0 & 1 & 1 & 0 & 0 \\ 0 & 1 & 1 & 0 & 0 \\ 0 & 0 & 0 & 1 & 0 \\ 1 & 0 & 0 & 0 & 1 \end{matrix}$$

(continued)

Example 2.15 (continued)
(as in the Example 2.10), so the indistinguishability classes with respect to Ψ are $\{b_1, b_5\}$, $\{b_2, b_3\}$, $\{b_4\}$. Let Ψ_1, Ψ_2, Ψ_3, Ψ_4 be, respectively, the following state descriptions for $b_1, \ldots, b_6$ extending Ψ:

$$\begin{matrix}1&0&0&0&1&1\\0&1&1&0&0&0\\0&1&1&0&0&0\\0&0&0&1&0&0\\1&0&0&0&1&1\\1&0&0&0&1&1\end{matrix}\qquad\begin{matrix}1&0&0&0&1&0\\0&1&1&0&0&1\\0&1&1&0&0&1\\0&0&0&1&0&0\\1&0&0&0&1&0\\0&1&1&0&0&1\end{matrix}\qquad\begin{matrix}1&0&0&0&1&0\\0&1&1&0&0&0\\0&1&1&0&0&0\\0&0&0&1&0&1\\1&0&0&0&1&0\\0&0&0&1&0&1\end{matrix}\qquad\begin{matrix}1&0&0&0&1&0\\0&1&1&0&0&0\\0&1&1&0&0&1\\0&0&0&1&0&1\\1&0&0&0&1&1\\0&0&0&0&0&1\end{matrix}$$

Then

$$p(\Psi_1(b_1, \ldots, b_6)) = p(\Psi_2(b_1, \ldots, b_6))$$
$$\geq p(\Psi_3(b_1, \ldots, b_6)) \geq p(\Psi_4(b_1, \ldots, b_6))$$

since

- b_6 is indistinguishable from b_1 and b_5 with respect to Ψ_1,
- b_6 is indistinguishable from b_2 and b_3 with respect to Ψ_2,
- b_6 is indistinguishable from b_4 with respect to Ψ_3,
- b_6 is not indistinguishable from any of $b_1, \ldots, b_5$ with respect to Ψ_4.

Interpreting the language as in Example 2.10, this says that an agent using a probability function satisfying PI, upon finding two pairs of concurring building blocks and an odd one, would consider it more likely that a next building block is like those in one of the pairs than that it is like the odd one, and being like the odd one more likely than being of any given particular form different from those already encountered.

It is one of the significant results of polyadic inductive logic that any probability function satisfying Li with Sx also satisfies the Principle of Induction. For a proof of this result, a special class of probability functions (the $u^{\bar{p},L}$ function, see for example Paris and Vencovská 2015, Chapter 29) satisfying Sx was defined and studied. They are the only probability functions satisfying WIP and Li with Sx, and any probability function satisfying just Sx can in some way be expressed using these functions. The proof also involves a generalisation of the Muirhead inequality, see Paris and Vencovská (2009).

The unary version[9] of the above Principle of Induction is as follows

Unary Principle of Induction, UPI *Let L contain only unary predicates. Let*

$$\Theta(b_1, \ldots, b_m) = \bigwedge_{j=1}^{m} \alpha_{k_j}(b_j)$$

be a state description and assume that the atom α_s *occurs at least as many times amongst the* $\alpha_{k_1}, \ldots, \alpha_{k_m}$ *as the atom* α_t. *Then*

$$p\left(\alpha_s(b_{m+1}) \mid \bigwedge_{j=1}^{m} \alpha_{k_j}(b_j)\right) \geq p\left(\alpha_t(b_{m+1}) \mid \bigwedge_{j=1}^{m} \alpha_{k_j}(b_j)\right).$$

Any probability function satisfying Ax satisfies UPI. The proof of this result preceded the one for the polyadic case and it is based on the de Finetti representation theorem for exchangeable probability functions and on the fact that the de Finetti measure for a probability function satisfying Ax is symmetric in a way which allows the Muirhead inequality to be applied.

Example 2.16 Using the setup of Example 2.12, suppose our head teacher consults Phil about sentences involving just one binary relation

'x likes y in the new school uniform'.

Then provided Phil's probability function p satisfies PI, his advice upon learning the complete specifications of how 4 students like each other in the new uniform corresponding to the matrix

$$\begin{matrix} 0 & 0 & 0 & 0 \\ 0 & 0 & 0 & 0 \\ 0 & 0 & 0 & 0 \\ 1 & 1 & 1 & 1 \end{matrix}$$

(continued)

[9] This is sightly stronger than PI as above restricted to unary languages in that it includes the requirement that the conditional probability of an individual satisfying an atom not instantiated in the evidence (that is, adding an individual in a way that makes it distinguishable from all the original ones) does not depend on which atom it is. It is natural to include this in the unary context but much less so in the polyadic where extending a state description to a new individual in a way that makes it distinguishable from all the original ones can fracture the original indistinguishability classes, as in fact happens with Ψ_4 extending Ψ in the above example.

Example 2.16 (continued)
would be that the behaviour of the next student corresponds with (not-necessarily-strictly) decreasing probability to the following matrices:

$$\begin{array}{ccccc} 0&0&0&0&0\\ 0&0&0&0&0\\ 0&0&0&0&0\\ 1&1&1&1&1\\ 0&0&0&0&0 \end{array} \quad \begin{array}{ccccc} 0&0&0&0&0\\ 0&0&0&0&0\\ 0&0&0&0&0\\ 1&1&1&1&1\\ 1&1&1&1&1 \end{array} \quad \begin{array}{ccccc} 0&0&0&0&a\\ 0&0&0&0&b\\ 0&0&0&0&c\\ 1&1&1&1&d\\ e&f&g&h&k \end{array}$$

where $a, b, c, d, e, f, g, h, k$ in the last matrix stands for any other assignment of 0,1 to the entries. That is, upon learning that three students do not like anybody in the new uniform and the fourth one likes everybody, Phil advises to expect that the next student is most likely not to like anybody in it and be liked just by the one student who liked everybody before, less likely to like everybody in it whilst being liked just by himself and the student who liked everybody before and least likely to display any other preferences. Note that if Phil's information corresponded to the other matrix from Example 2.12,

$$\begin{array}{cccc} 0&1&1&1\\ 1&1&1&1\\ 1&1&1&1\\ 1&1&1&1 \end{array}$$

then Phil's advice would have to be that the behaviour of the next student corresponds with (not-necessarily-strictly) decreasing probability to the following matrices:

$$\begin{array}{ccccc} 0&1&1&1&1\\ 1&1&1&1&1\\ 1&1&1&1&1\\ 1&1&1&1&1\\ 1&1&1&1&1 \end{array}, \quad \begin{array}{ccccc} 0&1&1&1&0\\ 1&1&1&1&1\\ 1&1&1&1&1\\ 1&1&1&1&1\\ 0&1&1&1&0 \end{array} \quad \begin{array}{ccccc} 0&1&1&1&a\\ 1&1&1&1&b\\ 1&1&1&1&c\\ 1&1&1&1&d\\ e&f&g&h&k \end{array},$$

that is, the second most probable option would be that the new student and the one who disliked just himself in the new uniform dislike each other and their own selves in it whilst liking and being liked by anybody else, and whilst anybody else likes everybody in it.

2.4 Conclusion

In this contribution, we have described the setup for pure inductive logic and discussed some of its principles. We illustrated their effect on the behaviour of hypothetical virtual agent who could be assumed to possess exclusively the information that is spelled out, and no other, and who aspires to be rational. The aim of this contribution was to bring more attention to this somewhat isolated area of study of rationality and to the extensive mathematical background that has been developed to support it.

References

de Finetti, B. 1974. *Theory of Probability*. vol. 1. New York: Wiley.
Gaifman, H. 1964. Concerning Measures on First Order Calculi. *Israel Journal of Mathematics* 2: 1–18.
Hajek, A., and N. Hall. 2002. Induction and Probability. In *The Blackwell Guide to the Philosophy of Science*, ed. P. Machamer, R. Silberstein, 149–172. Hoboken: Blackwell.
Johnson, W.E. 1932. Probability: The Deductive and Inductive Problems. *Mind* 41: 409–423.
Joyce, J. 2009. Accuracy and Coherence: Prospects for an Alethic Epistemology of Partial Belief. In *Degrees of Belief*, ed. F. Huber, C. Schmidt-Petri, 263–300. Berlin: Springer.
Paris, J.B., and A. Vencovská. 2009. Generalization of Muirhead's Inequality. *Journal of Mathematical Inequalities* 3 (2): 181–187.
Paris, J.B., and A. Vencovská. 2011. Symmetry's End? *Erkenntnis* 74 (1): 53–67.
Paris, J.B., and A. Vencovská. 2015. *Pure Inductive Logic*. Cambridge: Cambridge University Press.
Paris, J.B., and A. Vencovská. 2016. Combining Analogical Support in Pure Inductive Logic. *Erkenntnis* 82 (2): 401–419.
Paris, J.B., and A. Vencovská. 2019. Translation Invariance and Miller's Weather Example. *Journal of Logic, Language and Information* 28 (4): 489–514.
Paris, J.B., and A. Vencovská. Symmetry Principles in Pure Inductive Logic. *Logica*. Rickmansworth: College Publications.
Paris, J.B., and A. Vencovská. On the Strongest Principles of Rational Belief Assignment. To appear in *The Journal of Logic, Language and Information*.
Pettigrew, R. 2011. Epistemic Utility Arguments for Probabilism. In *Stanford Encyclopedia of Philosophy*, ed. E. Zalta. https://plato.stanford.edu/archIves/sum2020/entries/epistemic-utility/
Ronel, T., and A. Vencovská. 2014. Invariance principles in polyadic inductive logic. *Logique et Analyse* 57 (228): 541–561.
Savage, L.J. 1954. *The Foundations of Statistics*. New York: Wiley.

Chapter 3
Where Do We Stand on Maximal Entropy?

Jon Williamson

Abstract Edwin Jaynes' *principle of maximum entropy* holds that one should use the probability distribution with maximum entropy, from all those that fit the evidence, to draw inferences, because that is the distribution that is maximally non-committal with respect to propositions that are underdetermined by the evidence. The principle was widely applied in the years following its introduction in 1957, and in 1978 Jaynes took stock, writing the paper 'Where do we stand on maximum entropy?' to present his view of the state of the art. Jaynes' principle needs to be generalised to a principle of *maximal* entropy if it is to be applied to first-order inductive logic, where there may be no unique maximum entropy function. The development of this objective Bayesian inductive logic has also been very fertile and it is the task of this chapter to take stock. The chapter provides an introduction to the logic and its motivation, explaining how it overcomes some problems with Carnap's approach to inductive logic and with the subjective Bayesian approach. It also describes a range of recent results that shed light on features of the logic, its robustness and its decidability, as well as methods for performing inference in the logic.

Keywords Maximal entropy · Maximum entropy · Inductive logic · Probabilistic logic · Objective Bayesianism · Objective Bayesian inductive logic · Entropy-limit conjecture · Bayesian network

---

J. Williamson (✉)
Department of Philosophy, University of Manchester, Manchester, UK
e-mail: jon.williamson@manchester.ac.uk

H. Hosni, J. Landes (eds.), *Perspectives on Logics for Data-driven Reasoning*,
Logic, Argumentation & Reasoning 35,
https://doi.org/10.1007/978-3-031-77892-6_3

3.1 Introduction

In his pioneering work on information theory, Claude Shannon (1948, §6) argued that the amount of information carried by a discrete probability distribution should be measured by its *entropy*,

$$H(P) = -\sum_{\omega \in \Omega} P(\omega) \log P(\omega),$$

where Ω is a finite set of basic outcomes. In 1957, Edwin Jaynes put forward his 'principle of maximum entropy':

> [I]n making inferences on the basis of partial information we must use that probability distribution which has maximum entropy subject to whatever is known. This is the only unbiased assignment we can make; to use any other would amount to arbitrary assumption of information which by hypothesis we do not have. ... The maximum-entropy distribution may be asserted for the positive reason that it is uniquely determined as the one which is maximally noncommittal with regard to missing information. (Jaynes 1957, p. 623.)

Jaynes' principle of maximum entropy, or 'maxent' for short, was quickly taken up in areas such as physics, engineering and statistics and theoretical developments were rapid. So much so that at a conference on the Maximum Entropy Formalism, held at MIT in 1978, Jaynes decided that the time was right to take stock and present the state of the art in a paper entitled 'Where do we stand on maximum entropy?' (Jaynes 1979). In that paper, Jaynes presented the historical background to maxent and its key features, speculated about its future and discussed its application to irreversible statistical mechanics. Maxent has continued to be fruitfully applied to the sciences, and 2024 saw the 43rd International Conference on Bayesian and Maximum Entropy methods in Science and Engineering.

Jaynes argued that maxent can underpin a version of objective Bayesianism, which he viewed as providing an inductive 'logic of science' (Jaynes 2003).[1] In this chapter, we shall see that maxent and objective Bayesianism can indeed yield a viable inductive logic. This logic is called *objective Bayesian inductive logic*, or OBIL for short. The application of maxent to inductive logic has been an active area of collaborative research since the main ideas were set out in Williamson (2008) and Barnett and Paris (2008), and the time is now ripe to take stock and present the state of the art.

Section 3.2 presents an introduction to objective Bayesianism (Sect. 3.2.1) and its connection to inductive logic (Sect. 3.2.2). Section 3.3 highlights a number of special cases that are particularly well understood. In Sect. 3.4, we turn to the question of how to motivate OBIL, showing that it can be justified on the grounds that if one were to use an inductive logic to decide how to bet, OBIL would be needed to avoid incurring avoidable losses. Next, we compare OBIL to some

[1] See Rosenkrantz (1977) for an early philosophical account of the role of maxent in an objective Bayesian approach to inductive inference.

alternative approaches to inductive logic: Carnap's programme (Sect. 3.5) and two subjective Bayesian approaches (Sect. 3.6). We then consider the question of how to perform inference in the logic, in Sect. 3.7. We see that there is a large class of entailment relationships in OBIL that are decidable (Sect. 3.7.1), and inference can be performed by using augmented truth tables (Sect. 3.7.2) or probabilistic graphical models (Sect. 3.7.3), for example. We give a flavour of the logical properties of OBIL in Sect. 3.8 and discuss the question of language relativity in Sect. 3.9. In Sect. 3.10 we explore the connection between OBIL and a closely related approach, namely the 'entropy-limit' approach of Barnett and Paris (2008). Conclusions are drawn in Sect. 3.11.

This chapter will present key results without proof, so that the reader can quickly ascertain where we currently stand. While OBIL is now a mature theory, we shall see along the way that there remain many open questions, making it a potentially fruitful formalism for further research.

3.2 Objective Bayesianism and Inductive Logic

3.2.1 Objective Bayesianism

According to objective Bayesianism, the strengths of one's beliefs need to satisfy three kinds of norm to qualify as rational (Williamson 2010):[2]

Structural. Degrees of belief should satisfy the laws of probability. For example, the degree to which one believes that the next card to be drawn from a stack of standard playing cards will be black should be the sum of the degree to which one believes that it will be a spade and the degree to which one believes it will be a club, given that spades and clubs are the black suits.

Evidential. Degrees of belief should satisfy constraints imposed by evidence. In particular, they should be calibrated to empirical probabilities, insofar as one has evidence of these empirical probabilities. For example, if one establishes that the stack of playing cards is a complete deck, one should believe to degree 0.25 that the next card drawn is a club.

Equivocation. Degrees of belief should otherwise equivocate as far as possible between outcomes. For example, if one knows only that an experiment has four possible (mutually exclusive) outcomes and that outcome 1 has empirical probability 0.1, then one should believe that the next outcome will be outcome 2 to degree 0.3.[3]

[2] Note that this view of objective Bayesianism is very different to the version developed by Jaynes (2003). In particular, Jaynes rejected the idea of empirical probabilities, to which this version of objective Bayesianism appeals.

[3] By the Evidential norm, one should believe that it will be outcome 1 to degree 0.1. By the Structural norm, one should believe that it will be one of the remaining outcomes to degree 0.9.

Maxent provides a natural way to explicate the Equivocation norm. Entropy can be interpreted as the measure of the extent to which a probability function equivocates between the basic possibilities. Hence, the Equivocation norm can be implemented by selecting the probability function that has maximum entropy, from all those that satisfy constraints imposed by the Evidential norm.

From a technical point of view, this works rather straightforwardly when the space Ω of basic possibilities is finite. From a philosophical point of view, however, it is open to the charge that the choice of Ω may be rather arbitrary and yet may affect the inferences that are drawn, undermining the purported objectivity of objective Bayesianism. To address this philosophical objection, the approach taken by Williamson (2010, p. 156) is to take Ω to be the set of basic possibilities expressible in one's language:[4] that inferences depend upon the underlying language in this way is relatively unproblematic, because languages evolve to represent and reason about the world efficiently, and thus can be thought of as providing implicit evidence about the world. The difficulty here is that languages tend to be very rich. If one's language could be modelled by a finite propositional language, then it would induce a finite set Ω of all states $\pm a_1 \wedge \cdots \wedge \pm a_n$ of the atomic propositions $a_1, \ldots, a_n$ of the language, and again it would be straightforward to maximise entropy (see, e.g., Paris 1994). But more typically, one needs at least the richness of a first-order predicate language to express many of the propositions that feature in our inferences. A first-order predicate language does not induce a finite set of states, and it is no longer obvious how to maximise entropy. Hence there is a danger that this appeal to language mitigates a philosophical problem at the expense of introducing a technical problem.

Objective Bayesian inductive logic allows us to address this technical problem, however, as we shall now see.

3.2.2 Objective Bayesian Inductive Logic

OBIL considers inductive entailment relationships of the form:

$$\varphi_1^{X_1}, \ldots, \varphi_k^{X_k} \mathrel{|\!\!\!\approx} \psi^Y$$

Here, $\varphi_1, \ldots, \varphi_k, \psi$ are sentences of a first-order predicate language $\mathcal{L}$ and $X_1, \ldots, X_k, Y$ are sets of probabilities. The entailment relationship can be read: if $P(\varphi_1) \in X_1, \ldots$, and $P(\varphi_k) \in X_k$ then $P(\psi) \in Y$, for any rational belief function P. The premisses on the left-hand side of the entailment relation are

By Equivocation, one should assign each of these remaining outcomes the same probability, 0.3, in the absence of any other relevant evidence.

[4] This line of response is motivated by the suggestion of Keynes (1921) that one should only equivocate between indivisible possibilities.

interpreted as all the constraints on rational degrees of belief imposed by the Evidential norm, while the conclusion on the right-hand side follows just when each maximally equivocal probability function P, from all those that satisfy the premisses, also satisfies the conclusion. The key task is to say what constitutes 'maximally equivocal'.

First, we need to specify the framework more precisely. Here $\mathcal{L}$ is a pure first-order predicate language: it has relation symbols $U_1, \ldots, U_l$, constant symbols $t_1, t_2, \ldots$ and variable symbols $x_1, x_2, \ldots$, but no function symbols or equality. Sentences θ, φ_i, ψ etc. are formed in the usual way using quantifiers $\forall, \exists$, and connectives $\neg, \wedge, \vee, \rightarrow, \leftrightarrow$. Atomic sentences $a_1, a_2, \ldots$ are ordered so that those involving constants $t_1, \ldots, t_n$ occur in the ordering before those involving t_{n+1}.[5] We shall consider the sublanguages $\mathcal{L}_n$ that have all the syntactic apparatus of $\mathcal{L}$ but involve only the constants $t_1, \ldots, t_n$. The *n-states* $\omega \in \Omega_n$ of $\mathcal{L}$ are the states of $\mathcal{L}_n$, i.e., the sentences of the form $\pm a_1 \wedge \cdots \wedge \pm a_{r_n}$, where $a_1, \ldots, a_{r_n}$ are the atomic sentences of $\mathcal{L}_n$.

We shall take $X_1, \ldots, X_k, Y$ to be intervals. This is because constraints imposed by the Evidential norm are convex: if the empirical probability of φ is known to be either x or y, where $y > x$, but it is not known which, then the Evidential norm deems any value in the interval $[x, y]$ to be an admissible degree of belief (Williamson 2010, §3.3). We shall abbreviate the trivial interval $[x, x]$ by x. We shall also abbreviate a premiss or conclusion statement of the form $\varphi^{[1,1]}$ by the categorical (i.e., unqualified) sentence φ.

A probability function P is a function defined on the sentences of $\mathcal{L}$ such that:

P1. If τ is a deductive tautology, i.e., $\models \tau$, then $P(\tau) = 1$.
P2. If θ and φ are mutually exclusive, i.e., $\models \neg(\theta \wedge \varphi)$, then $P(\theta \vee \varphi) = P(\theta) + P(\varphi)$.
P3. $P(\exists x \theta(x)) = \sup_m P\left(\bigvee_{i=1}^m \theta(t_i)\right)$.

Axiom P3, which is sometimes called *Gaifman's condition*, presupposes that each member of the domain of discourse is named by some constant symbol t_i. A probability function is uniquely determined by its values on the n-states (Williamson 2017, Chapter 2). The set of all probability functions on $\mathcal{L}$ is denoted by $\mathbb{P}$. We shall be particularly interested in the set of probability functions that satisfy the evidential constraints:

$$\mathbb{E} \stackrel{\text{df}}{=} [\varphi_1^{X_1}, \ldots, \varphi_k^{X_k}] \stackrel{\text{df}}{=} \{P \in \mathbb{P} : P(\varphi_1) \in X_1, \ldots, P(\varphi_k) \in X_k\}.$$

Now we are in a position to see what constitutes 'maximally equivocal'. We define the n-entropy:

$$H_n(P) \stackrel{\text{df}}{=} -\sum_{\omega \in \Omega_n} P(\omega) \log P(\omega).$$

[5] See Sect. 3.9 for results that indicate that OBIL is invariant under the precise ordering.

We then say that P has greater entropy than Q iff the n-entropy of P eventually dominates that of Q, i.e., iff there is an $N \in \mathbb{N}$ such that for all $n \geq N$, $H_n(P) > H_n(Q)$. The greater-entropy relation yields a partial ordering of probability functions, which may contain maximal elements (undominated functions) but need not necessarily contain maximum elements (functions that dominate all others). We thus define the maximally equivocal functions in $\mathbb{E}$ to be those with maximal entropy:

$$\text{maxent}\,\mathbb{E} \stackrel{\text{df}}{=} \{P \in \mathbb{E} : \text{ there is no } Q \in \mathbb{E} \text{ that has greater entropy than } P\}.$$

This yields what we might call the 'principle of maximal entropy'—an extension of the principle of maximum entropy to the setting of an infinite predicate language:

Maximal Entropy Principle. In making inferences on the basis of partial information we must use the probability distributions which have maximal entropy, among all those that satisfy the evidential constraints.

We can then use the maximal entropy principle to provide semantics for OBIL:

$$\varphi_1^{X_1}, \ldots, \varphi_k^{X_k} \mathrel{|\!\!\approx^{\circ}} \psi^Y \text{ iff } P(\psi) \in Y \text{ for all } P \in \text{maxent}\,\mathbb{E},$$

as long as $\text{maxent}\,\mathbb{E} \neq \emptyset$. There are two cases in which $\text{maxent}\,\mathbb{E} = \emptyset$, and we need some conventions to cover these cases.[6] The first is where the premisses are unsatisfiable, $\mathbb{E} = \emptyset$. In that case, it is desirable to avoid 'explosion', i.e., the phenomenon that any conclusion follows, because it is never rational to believe everything. We thus consider $\text{maxent}\,\mathbb{P}$ instead of $\text{maxent}\,\mathbb{E}$ when $\mathbb{E} = \emptyset$. $\mathbb{P}$ has a unique maximiser, namely the *equivocator function* $P_=$ defined by $P_=(\omega) = 1/|\Omega_n| = 1/2^{r_n}$ for all $\omega \in \Omega_n$ and $n \geq 1$. Hence an entailment relationship with inconsistent premisses holds iff $P_=(\psi) \in Y$. $P_=(\psi)$ is called the *measure* of ψ.[7] The second case occurs where $\mathbb{E} \neq \emptyset$ but for any probability function in $\mathbb{E}$ there is another function with greater entropy. In this case, we shall consider $\mathbb{E}$ instead of $\text{maxent}\,\mathbb{E}$, so the entailment relationship holds whenever $P(\psi) \in Y$ for all $P \in \mathbb{E}$.

If there are no premisses, the equivocator function is used for inference. If $\mathrel{|\!\!\approx^{\circ}} \psi$ then ψ is said to be an *inductive tautology*. Equivalently, $P_=(\psi) = 1$, i.e., ψ has measure 1. If $\mathrel{|\!\!\approx^{\circ}} \neg\psi$ (i.e., ψ has measure 0) then ψ is an *inductive contradiction*. If $\not\mathrel{|\!\!\approx^{\circ}} \neg\psi$ (i.e., ψ has positive measure) then ψ is *inductively consistent*. Similarly, if $\not\mathrel{|\!\!\approx^{\circ}} \neg(\psi \wedge \theta)$ (i.e., $P_=(\psi \wedge \theta) > 0$) then ψ is *inductively consistent* with θ. If $\mathrel{|\!\!\approx^{\circ}} \psi \leftrightarrow \theta$ then ψ and θ are *inductively equivalent*.

---

6 These cases will not be pursued further in this chapter, but see Williamson (2010, 2017) for further discussion and alternative conventions.

7 The equivocator function is an analogue of Lebesgue measure, via a bijective mapping between probability functions on $\mathcal{L}$ and probability measures on a σ-field of subsets of the unit interval (Williamson 2017, §2.6). The restriction of $P_=$ to the sentences of any finite sublanguage $\mathcal{L}_n$ corresponds to the uniform distribution on the finite set of n-states Ω_n. There is no uniform distribution on $\mathcal{L}$ itself, because there are infinitely many n-states of $\mathcal{L}$, since $n = 1, 2, \ldots$.

3.3 Important Cases

This section will survey some important special cases in which there is a unique maximal entropy function and where this function can be determined rather straightforwardly. As we shall see in subsequent sections, these are the cases that have been explored in most depth.

3.3.1 $\mathbb{E}$ Has Finitely Generated Consequences

Recall that, given premisses $\varphi_1^{X_1}, \ldots, \varphi_k^{X_k}$, $\mathbb{E} \stackrel{\text{df}}{=} [\varphi_1^{X_1}, \ldots, \varphi_k^{X_k}] \stackrel{\text{df}}{=} \{P \in \mathbb{P} : P(\varphi_1) \in X_1, \ldots, P(\varphi_k) \in X_k\}$. One special case occurs when the consequences of these premisses can be characterised as the consequences of some set of quantifier-free premisses:

Definition 3.1 (Finitely Generated) $\mathbb{E}$ is *finitely generated* if there exist quantifier-free sentences $\theta_1, \ldots, \theta_j$ and intervals $Z_1, \ldots, Z_j$ such that $\mathbb{E} = [\theta_1^{Z_1}, \ldots, \theta_j^{Z_j}]$. $\mathbb{E}$ has *finitely generated consequences* if there exist quantifier-free sentences $\theta_1, \ldots, \theta_j$ and intervals $Z_1, \ldots, Z_j$ such that maxent $\mathbb{E} = \text{maxent}[\theta_1^{Z_1}, \ldots, \theta_j^{Z_j}]$. In this latter case, $\theta_1^{Z_1}, \ldots, \theta_j^{Z_j}$ are called *generating statements* for (the consequences of) $\mathbb{E}$.

Clearly if $\mathbb{E}$ is finitely generated then it has finitely generated consequences, and if the premisses are themselves all quantifier-free then $\mathbb{E}$ is finitely generated. But even when the premisses are not quantifier-free, $\mathbb{E}$ often turns out to have finitely generated consequences. To see when this is so, we require a definition from Landes et al. (2024, §4):

Definition 3.2 (Support) Suppose $a_{i_1}, \ldots, a_{i_m}$ include all the atomic propositions that appear in sentence φ of $\mathcal{L}$, and let $\Xi_\varphi \stackrel{\text{df}}{=} \{\pm a_{i_1} \wedge \ldots \wedge \pm a_{i_m}\}$ be the set of states of these atomic propositions. If φ contains no atomic propositions, we take $\Xi_\varphi \stackrel{\text{df}}{=} \{a_1, \neg a_1\}$. The *support* $\check{\varphi}$ of φ is the disjunction of states in Ξ_φ that are inductively consistent with φ,

$$\check{\varphi} \stackrel{\text{df}}{=} \bigvee\{\xi \in \Xi_\varphi : P_{=}(\xi \wedge \varphi) > 0\}.$$

For example, if $\varphi = \exists x(Ut_1 \wedge Vt_1x)$ then $\check{\varphi} = Ut_1$.

Definition 3.3 (Support-Satisfiable) Let $\check{\mathbb{E}} \stackrel{\text{df}}{=} [\check{\varphi}_1^{X_1}, \ldots, \check{\varphi}_k^{X_k}]$. We say that the premisses $\varphi_1^{X_1}, \ldots, \varphi_k^{X_k}$ have *satisfiable support*, or that $\mathbb{E}$ is *support-satisfiable*, if $\check{\mathbb{E}} \neq \emptyset$. An entailment relationship is *support-satisfiable* if $\mathbb{E}$ is support-satisfiable.

$\mathbb{E} = [\exists x Vt_1x, (Ut_2 \vee \forall x Rx)^{0.9}, Ut_1 \to Vt_1t_3, (Ut_1 \vee (\exists x Vxt_3 \to Ut_2))^{[0.95,1]}]$, for example, is support-satisfiable, as we see in Sect. 3.7. Support-satisfiability is violated in cases where the premisses force an inductive tautology to have probability greater than 0 or force an inductive contradiction to have probability less than 1 (Landes et al. 2024, §10.2). $\mathbb{E} = [\forall x U x^{0.7}]$, for instance, is not support-satisfiable, because the constraint $\forall x U x^{0.7}$ forces positive probability on the measure-zero sentence $\forall x U x$. Note that if $\mathbb{E}$ is non-empty and finitely generated then it is support-satisfiable. Landes et al. (2024, §5) show that:

Theorem 3.1 *If* $\mathbb{E}$ *is support-satisfiable then it has finitely generated consequences, with generating statements* $\check{\varphi}_1^{X_1}, \ldots, \check{\varphi}_k^{X_k}$.

We also have (see Landes et al. 2024):

Theorem 3.2 *If* $\mathbb{E}$ *is closed and has finitely generated consequences then it contains a unique maximal entropy function* $P^\dagger$, *i.e.,* $\text{maxent}\, \mathbb{E} = \{P^\dagger\}$.

Moreover, $P^\dagger$ can be characterised as follows (Landes et al. 2024, §3.3). For any n, let $P_n^\dagger$ be the n-entropy maximiser, $P_n^\dagger \stackrel{\text{df}}{=} \arg\max_{P \in \mathbb{E}} H_n(P)$, if it exists. (Since $\mathbb{E}$ is convex and n-entropy is strictly concave, there can be no more than one n-entropy maximiser.) Now, if $\mathbb{E}$ is closed then $P_n^\dagger$ does indeed exist for each n. Consider any n large enough that all the quantifier-free generating statements for $\mathbb{E}$ are expressible in $\mathcal{L}_n$. Then the entropy maximiser $P^\dagger$ is the probability function that agrees with $P_n^\dagger$ on $\mathcal{L}_n$ but equivocates elsewhere: $P^\dagger$ is defined by $P^\dagger(\omega_m) = P_n^\dagger(\omega_n) P_=(\zeta)$ for each $m \geq n$ and m-state $\omega_m = \omega_n \wedge \zeta$ where $\omega_n \in \Omega_n$.

We thus have a sequence of increasingly general situations in which there is guaranteed to be a unique maximal entropy function:

$\mathbb{E}$ is closed, non-empty and finitely generated;
$\Rightarrow$ $\mathbb{E}$ is closed and support-satisfiable;
$\Rightarrow$ $\mathbb{E}$ is closed and has finitely generated consequences.

3.3.2 $\mathbb{E}$ Is Closed in Entropy

It turns out that we can relax both conditions of Theorem 3.2 by considering the convergence of the n-entropy maximisers.

Definition 3.4 (Limit in Entropy) $P \in \mathbb{P}$ is a *limit in entropy* of $\mathbb{E}$ if $|H_n(P_n^\dagger) - H_n(P)| \longrightarrow 0$ as $n \longrightarrow \infty$. $\mathbb{E}$ is *closed in entropy* if it contains some limit in entropy.

For example, $\mathbb{E} = [\forall x U x^{0.7}]$ is closed in entropy but $\mathbb{E} = [U t_1 \vee \exists x \forall y V x y]$ is not. Landes et al. (2021, §5) provide several tests that can help to determine whether $\mathbb{E}$ is closed in entropy. The concept of closure in entropy is useful because (Landes et al. 2023, Theorem 16):

Theorem 3.3 *If* $\mathbb{E}$ *is closed in entropy then* $\text{maxent}\, \mathbb{E} = \{P^\dagger\}$ *where* $P^\dagger$ *is the unique limit in entropy of* $\mathbb{E}$.

Thus $P^\dagger$ can be determined from the behaviour of the $P_n^\dagger$ as n increases.

Let us consider an example of a case in which $\mathbb{E}$ does not have finitely generated consequences (Landes et al. 2023, Example 17):

$$\forall x U x^{0.7} \mathrel{|\!\!\!\approx^{\circ}} U t_1^{0.85}$$

Here, $\mathbb{E}$ contains the following limit in entropy:

$$P(\omega) = \begin{cases} 0.7 + \frac{0.3}{2^n} & : \quad \omega = U t_1 \wedge \ldots \wedge U t_n \\ \frac{0.3}{2^n} & : \quad \omega \models \neg(U t_1 \wedge \ldots \wedge U t_n) \end{cases} .$$

Hence, maxent $\mathbb{E} = \{P\}$ and $P(Ut_1) = 0.7 + (0.3/2) = 0.85$. Thus the above entailment relationship does indeed hold.

3.4 Motivation

The norms of objective Bayesianism can be justified by appeal to the following considerations: the strengths of one's beliefs guide one's actions; in turn, one's actions can expose one to potential loss; one should not adopt beliefs that expose one to avoidable losses.

The Structural norm is the claim that rational degrees of belief are probabilities. This can be justified on the grounds that one should not adopt beliefs that expose one to *guaranteed* loss, however events turn out. In the finite case this is the well known Dutch Book argument of Ramsey (1926) and de Finetti (1937). This Dutch Book argument can be extended to the situation in which degrees of belief are defined over sentences of a first-order predicate language. Suppose that the loss incurred by believing sentence θ to degree x is:

$$L(\theta, x) = (x - I_\theta) S_\theta,$$

where $S_\theta \in \mathbb{R}$ is an unknown stake (positive or negative) and the indicator function I_θ takes the value 1 if θ is true and 0 if θ is false. A *Dutch book* on set Θ of sentences is a combination of stakes $S_\theta \in \mathbb{R}$ for each $\theta \in \Theta$ that guarantees some fixed loss, i.e., that ensures a positive finite loss $L(\Theta) \stackrel{\text{df}}{=} \sum_{\theta \in \Theta} L(\theta, x) \in (0, \infty)$ whichever sentences $\theta \in \Theta$ turn out to be true. Then (Williamson 2017, Theorem 9.1):

Theorem 3.4 (Dutch Book Theorem) *Degrees of belief on sentences of $\mathcal{L}$ avoid the possibility of a Dutch book if and only if they satisfy the axioms of probability.*

The Evidential and Equivocation norms are explicated by means of the maximal entropy principle, and this principle can be justified on the grounds that one should not adopt beliefs that expose one to positive *expected* loss, with respect to a default loss function. A default loss function represents the losses one might reasonably anticipate, in the absence of any information about any particular decision scenario and hence the true losses to which one will be exposed. Let $L_\pi(\theta, P)$ signify the loss incurred by adopting belief function P when $\theta \in \pi$ turns out to be true, where π is a partition of sentences. For partitions π_1, π_2, π, write $\pi = \pi_1 \times \pi_2$ when for each $\theta_1 \in \pi_1, \theta_2 \in \pi_2$ there is some $\theta \in \pi$ such that $\theta \equiv \theta_1 \wedge \theta_2$. A probability function P renders π_1 and π_2 independent, written $\pi_1 \perp\!\!\!\perp_P \pi_2$, when $P(\theta_1 \wedge \theta_2) = P(\theta_1)P(\theta_2)$ for each $\theta_1 \in \pi_1, \theta_2 \in \pi_2$. Consider the following constraints on a default loss function:

L1. No loss is incurred when one fully believes the sentence that turns out to be true: $L_\pi(\theta, P) = 0$ if $P(\theta) = 1$.
L2. Loss $L_\pi(\theta, P)$ strictly increases as $P(\theta)$ decreases from 1 to 0.
L3. Loss $L_\pi(\theta, P)$ depends only on $P(\theta)$, not on $P(\varphi)$ for other partition members φ.
L4. Losses are additive over independent partitions: if $\pi = \pi_1 \times \pi_2$ where $\pi_1 \perp\!\!\!\perp_P \pi_2$, then for each $\theta \in \pi$, $L_\pi(\theta, P) = L_{\pi_1}(\theta_1, P) + L_{\pi_2}(\theta_2, P)$, where $\theta_1 \in \pi_1, \theta_2 \in \pi_2$ are such that $\theta \equiv \theta_1 \wedge \theta_2$.
L5. Loss $L_\pi(\theta, P)$ should not depend on the partition π in which θ occurs: there is some function L such that $L_\pi(\theta, P) = L(\theta, P)$ for all partitions π in which θ occurs.

These desiderata are enough to ensure that the default loss function is logarithmic (Williamson 2017, Theorem 9.2):

Theorem 3.5 *L1–L5 imply that* $L(\theta, P) = -k \log P(\theta)$, *for some constant* $k > 0$.

Let P^* be the empirical probability function. Then we can consider the expected loss to which one is exposed by beliefs on $\mathcal{L}_n$, for $n \geq 1$, when one adopts $P \in \mathbb{P}$ as one's belief function:

$$S_n(P^*, P) = \sum_{\omega \in \Omega_n} P^*(\omega) L(\omega, P)$$
$$= -k \sum_{\omega \in \Omega_n} P^*(\omega) \log P(\omega).$$

Now suppose that evidence establishes that the empirical probability function P^* lies in a non-empty, closed convex set $\mathbb{E}$ of probability functions. On $\mathcal{L}$ as a whole, $P \in \mathbb{P}$ has lower worst-case expected loss than $Q \in \mathbb{P}$ if there is some N such that for all $n \geq N$,

$$\sup_{P^* \in \mathbb{E}} S_n(P^*, P) < \sup_{P^* \in \mathbb{E}} S_n(P^*, Q).$$

Generalising a result of Topsøe (1979), it turns out that (Williamson 2017, Theorem 9.3):

Theorem 3.6 (Minimax) *If* $\mathbb{E}$ *is finitely generated then there is a unique probability function that has minimal worst-case expected loss and this is the function* $P^\dagger \in \mathbb{E}$ *that has maximal entropy.*

This result thus provides a justification for the maximal entropy principle, at least in an important special case. The main open question here concerns how far one can relax the assumption that $\mathbb{E}$ is finitely generated.

It is worth noting that while this result can be viewed as providing a pragmatic justification of the maximal entropy principle, it can be recast as an epistemic justification by reinterpreting the function L_n as a measure of the epistemic inaccuracy of P, instead of as a default loss function (Williamson 2018). Note too that the line of motivation presented in this section appeals to two arguments: a Dutch Book theorem to justify the claim that a rational belief function is a probability function and the Minimax theorem to justify the maximal entropy principle. Landes and Williamson (2013, 2015) explore the possibility of a unified argument that seeks to justify all the norms of OBIL at once.

3.5 Relation to Carnap's Programme

In this section, we briefly consider how OBIL compares to Rudolf Carnap's well known system of inductive logic (Carnap 1952).

Carnap also considered inductive logic defined on a first-order predicate language, but in the special case in which the premisses are all categorical, i.e., no uncertainty attaches to the premiss sentences. Carnap's approach can be thought of as providing semantics for entailment relationships of the form:

$$\varphi_1, \ldots, \varphi_k \approx_\lambda \psi^Y,$$

where $\lambda \in [0, \infty]$ is a parameter that selects one out of a continuum of entailment relations. Carnap was particularly interested in inference from a sample $\pm Ut_1, \ldots, \pm Ut_k$ of past observations. For Carnap, an entailment relationship of the form

$$\pm Ut_1, \ldots, \pm Ut_k \approx_\lambda Ut_{k+1}^Y$$

holds just when

$$c_\lambda(Ut_{k+1} | \pm Ut_1, \ldots, \pm Ut_k) \stackrel{\text{df}}{=} \frac{\#U + \lambda/2}{k + \lambda} \in Y,$$

where $\#U$ is the number of positive instances of U in the sample.[8]

There are a number of key differences between OBIL and Carnap's approach. Firstly, OBIL is more general, as it is not restricted to categorical premisses.

[8] Here, c_λ is one of a continuum of probability functions that is defined by the above assignment of conditional probabilities.

Because Carnap's approach appeals to conditional probabilities, the premisses that are conditioned on must be categorical. Second, Carnap's approach yields a continuum of inductive logics, and, unfortunately, he gave little concrete guidance as to which logic to choose from this continuum. Thus, inductive logic is somewhat underdetermined in Carnap's framework.

The third and perhaps the most fundamental difference is that Carnap's approach tries to embed learning from experience within the logic, while OBIL takes this to be a separate, statistical question. The extent to which Carnap's logics exhibit learning from experience varies with the parameter λ. If $\lambda = 0$, then the probability of the next outcome being positive is set to the observed frequency $\#U/k$ of positive outcomes in the sample. As λ increases, the influence of the sample decreases until, when $\lambda = \infty$, the probability of the next outcome being positive remains at the value $\frac{1}{2}$ regardless of the proportion of positive outcomes in the sample.

OBIL, on the other hand, takes the question of the influence of the sample to have been determined when formulating the premisses. Recall that objective Bayesianism requires degrees of belief to be appropriately calibrated to empirical probabilities. The way this is fleshed out by Williamson (2017, Chapter 7) is by appeal to confidence intervals. Given a sample $\pm Ut_1, \ldots, \pm Ut_k$ of outcomes, consider the narrowest confidence interval $[l, u]$ within which one is prepared to infer that the empirical probability of a positive outcome lies. If there is no more pertinent information about the next outcome, the Evidential norm would say that one should believe to some degree in the interval $[l, u]$ that the next outcome will be positive. One thus adds $Ut_{k+1}{}^{[l,u]}$ to the list of premisses. This is because the premisses are supposed to represent all the relevant constraints imposed by the Evidential norm on degrees of belief. Therefore, the objective Bayesian needs to consider as premisses $\pm Ut_1, \ldots, \pm Ut_k, Ut_{k+1}{}^{[l,u]}, Ut_{k+2}{}^{[l,u]}, \ldots$. The maximal entropy principle would then yield, for example,

$$\pm Ut_1, \ldots, \pm Ut_k, Ut_{k+1}{}^{[.7,.8]}, Ut_{k+2}{}^{[.7,.8]}, \ldots \mathrel{|\!\!\approx} Ut_{k+1}{}^{.7}.$$

Thus, OBIL admits a clear demarcation between entailment in inductive logic and the 'knowledge representation' process of formulating suitable premisses, which can itself involve statistical inference. Theoretical work on OBIL often focusses on entailment, though it presupposes adequate knowledge representation. This demarcation accords well with the analogous case of deductive logic, where considerable work is required in formulating implicit as well as explicit premisses, but where the logical theory tends to presuppose that this work has been done and focusses on entailment. Here, constraints imposed by a sample, such as $Ut_{k+1}{}^{[.7,.8]}$, can be thought of as implicit premisses.

There are advantages to keeping these two aspects of induction distinct, rather than conflating them as Carnap does. As Williamson (2017, Chapter 4) argues, problems can arise for Carnap's approach because the probabilities are always treated as 'exchangeable', i.e., invariant under permutations of the constants. Exchangeability is sometimes appropriate (e.g., when drawing inferences from a sample from independent and identically distributed random variables), but not always. OBIL's separation of these statistical concerns from the logic allows for more flexibility with respect to statistical methods: exchangeability can be invoked where appropriate but need not always be adhered to.

3.6 Relation to Subjective Bayesianism

The subjective Bayesian approach to inductive inference has some points in common with Carnap's approach. In particular, it invokes conditional probabilities and thus requires categorical premisses. There are two key variants.

The *precise* subjectivist approach appeals to a prior probability function P_0 and holds that:

$$\varphi_1, \ldots, \varphi_k \approx_{P_0} \psi^Y \text{ iff } P_0(\psi|\varphi_1, \ldots, \varphi_k) \in Y.$$

Howson (2000) was a notable advocate of this sort of approach. It can be criticised as being radically underdetermined: according to the subjectivist any probability function is admissible as a prior probability function and there are no rational grounds for choosing one prior over another. Rather, the task is to elicit an agent's subjective degrees of belief and represent these using the prior probability function.

In order to address the problem of underdetermination, one could, of course, argue for stronger constraints on rational degrees of belief, but this arguably leads us back to objective Bayesianism. Alternatively one could adopt an *imprecise* subjectivist approach (Nilsson 1986):

$$\varphi_1, \ldots, \varphi_k \approx \psi^Y \text{ iff } P(\psi|\varphi_1, \ldots, \varphi_k) \in Y \text{ for all } P \in \mathbb{P}.$$

This can be generalised to what sometimes known as the *standard semantics* for inductive logic (Williamson 2017, §3.5):

$$\varphi_1^{X_1}, \ldots, \varphi_k^{X_k} \approx \psi^Y \text{ iff } P(\psi) \in Y \text{ for all } P \in \mathbb{E}.$$

Unfortunately, however, the imprecise approach arguably trades one problem (underdetermination) for another, namely weakness: it is often the case that the most one can infer from given premisses is that the probability of conclusion sentence ψ lies in the unit interval. OBIL usually motivates much stronger inferences.

A key point of difference between subjective and objective Bayesianism concerns conditional probabilities. Conditional probabilities are absolutely central to subjective Bayesianism, because evidence is taken into account by conditionalising, i.e., by ensuring that all probabilities are conditional on total evidence (see, e.g., Howson and Urbach 1989, §3.h). This stands in marked contrast to objective Bayesianism as developed above, which takes evidence into account by means of calibration to empirical probabilities and applying the maximal entropy principle.[9]

The question nevertheless remains as to the connection between the two ways of handling evidence. Landes et al. (2023, Theorem 34) show that updating in OBIL captures Bayesian conditionalisation as a special case:[10]

Theorem 3.7 (Bayesian Conditionalisation) *Given categorical premisses* $\varphi_1, \ldots, \varphi_k$, *if* $P_=(\varphi_1 \wedge \cdots \wedge \varphi_k) > 0$ *then*

$$\text{maxent}\,\mathbb{E} = \{P_=(\cdot|\varphi_1, \ldots, \varphi_k)\} = \{P_=(\cdot|\check{\varphi}_1, \ldots, \check{\varphi}_k)\}.$$

Landes et al. (2023, Theorem 41) go on to show that a generalisation of Bayesian conditionalisation, namely Jeffrey conditionalisation (Jeffrey 2004, §3.2), also holds in OBIL:

Theorem 3.8 (Jeffrey Conditionalisation) *Given* $c \in (0, 1)$ *and premiss* φ^c, *if* $P_=(\varphi) \in (0, 1)$ *then*

$$\begin{aligned}\text{maxent}\,\mathbb{E} &= \{c \cdot P_=(\cdot|\varphi) + (1-c) \cdot P_=(\cdot|\neg\varphi)\}\\ &= \{c \cdot P_=(\cdot|\check{\varphi}) + (1-c) \cdot P_=(\cdot|\neg\check{\varphi})\}.\end{aligned}$$

3.7 Inference

In this section, we turn to questions of inference in OBIL.

3.7.1 Decidability

As Landes et al. (2024, Corollary 1) observe, inference in inductive logic under the standard semantics is undecidable. However, there is a surprisingly large decidable class of entailment relationships in OBIL—decidable in the sense that there is an effective procedure for deciding whether a given entailment relationship lies in the class and, if so, whether the entailment is inductively valid. To obtain decidability results, we need to suppose that all intervals $X_1, \ldots, X_k, Y$ are expressible in the sense that they have endpoints that are expressed as terminating decimal fractions to some fixed number of decimal places. First, Landes et al. (2024, Theorem 1) show:

9 Indeed, Williamson (2023) argues that coherent calibration to empirical probabilities requires abandoning conditionalisation in favour of maximising entropy.

10 This extends a result of Seidenfeld (1986, §2.1) from the finite case to the case of a first-order predicate language.

Theorem 3.9 (Quantifier-Free Decidability) *The class of all quantifier-free entailment relationships is decidable in OBIL.*

Then, in virtue of the fact that there is an effective procedure for determining the support of any sentence, and this support is quantifier-free (Landes et al. 2024, Theorem 5),

Theorem 3.10 (Support-Satisfiable Decidability) *The class of all support-satisfiable entailment relationships is decidable in OBIL.*

3.7.2 Truth Tables

Given that such a large class of entailment relationships is decidable, the question arises as to how one might determine whether a given support-satisfiable entailment relationship is inductively valid. In this subsection we see that truth tables can be used for inference, while in the next, that probabilistic graphical models can be used for inference.

We shall illustrate these methods by providing a running example—see Landes et al. (2024) for more detail on the methods themselves. Consider the following entailment relationship:

$$\exists x Vt_1x, \;\; (Ut_2 \vee \forall x Rx)^{0.9}, \;\; Ut_1 \rightarrow Vt_1t_3, \;\; (Ut_1 \vee (\exists x Vxt_3 \rightarrow Ut_2))^{[0.95,1]} \;\; \mid\!\approx^{\circ} Vt_1t_3{}^{[.5,1]}$$

We can enumerate the atomic propositions as follows:

$$a_1 : Ut_1, \quad a_2 : Rt_1, \quad a_3 : Vt_1t_1,$$

$$a_4 : Ut_2, \quad a_5 : Rt_2, \quad a_6 : Vt_1t_2, \quad a_7 : Vt_2t_1, \quad a_8 : Vt_2t_2,$$

$$a_9 : Ut_3, \quad a_{10} : Rt_3, \quad a_{11} : Vt_1t_3, \quad a_{12} : Vt_3t_1, \quad a_{13} : Vt_2t_3, \quad \cdots$$

The support of each premiss sentence is:

i	φ_i	$\check{\varphi}_i$
1	$\exists x Vt_1x$	$a_1 \vee \neg a_1$
2	$Ut_2 \vee \forall x Rx$	a_4
3	$Ut_1 \rightarrow Vt_1t_3$	$a_1 \rightarrow a_{11}$
4	$Ut_1 \vee (\exists x Vxt_3 \rightarrow Ut_2)$	$a_1 \vee a_4$

Note in particular for φ_1, i.e., $\exists x Vt_1x$, we have that $\check{\varphi}_1$ is $a_1 \vee \neg a_1$ because φ_1 mentions no atomic propositions and $P_=(a_1 \wedge \exists x Vt_1x) = P_=(\neg a_1 \wedge \exists x Vt_1x) = 1/2 > 0$. Intuitively, since $\exists x Vt_1x$ is an inductive tautology, it is treated as if it were a deductive tautology when determining $\check{\varphi}_1$. In determining $\check{\varphi}_2$, the disjunct $\forall x Rx$ is an inductive contradiction and so ignored. Strictly speaking, $\check{\varphi}_3$ is defined as $(a_1 \wedge a_{11}) \vee (\neg a_1 \wedge a_{11}) \vee (\neg a_1 \wedge \neg a_{11})$ but we abbreviate this sentence by the logically equivalent sentence $a_1 \rightarrow a_{11}$. Similarly for $\check{\varphi}_4$. Note that $\exists x Vxt_3$ is an inductive tautology and hence treated as a deductive tautology when determining $\check{\varphi}_4$. The support of the conclusion is simply a_{11}.

The support inference is thus:

$$a_1 \vee \neg a_1, \quad a_4^{.9}, \quad a_1 \rightarrow a_{11}, \quad a_1 \vee a_4{}^{[0.95,1]} \mathrel{|\!\!\approx} a_{11}^{[0.5,1]}$$

The states of the atomic propositions that occur in the inference are $\Xi = \{\pm a_1 \wedge \pm a_4 \wedge \pm a_{11}\}$. These correspond to rows of the truth table for the support sentences:

$P^\dagger$	a_1	a_4	a_{11}	$a_1 \vee \neg a_1$	a_4	$a_1 \rightarrow a_{11}$	$a_1 \vee a_4$	a_{11}
0.3	T	T	T	T	T	T	T	T
0	T	T	F	T	T	F	T	F
0.05	T	F	T	T	F	T	T	T
0	T	F	F	T	F	F	T	F
0.3	F	T	T	T	T	T	T	T
0.3	F	T	F	T	T	T	T	F
0.025	F	F	T	T	F	T	F	T
0.025	F	F	F	T	F	T	F	F

The first column of the truth table shows the probability given to the state corresponding to each row of the truth table by the maximal entropy function $P^\dagger$. This is found by maximising the entropy $-\sum_{\xi \in \Xi} P(\xi) \log P(\xi)$ subject to the constraints imposed by the premisses. Standard numerical methods such as gradient ascent can be used to straightforwardly obtain the values that maximise entropy (see, e.g., Boyd and Vandenberghe 2004). The probability of each statement is the sum of the probabilities at the rows of the truth table at which it is true. We see then that $P^\dagger(a_{11}) = 0.3 + 0.05 + 0.3 + 0.025 = 0.675 \in [0.5, 1]$, so the entailment relationship is indeed inductively valid in OBIL.

3.7.3 Objective Bayesian Nets

The truth-table method is simple and powerful, but does not scale well to large and complex entailment relationships, because the number of rows in a truth table increases exponentially in the number of atomic sentences that feature in it. However, there is an alternative method for inference that is more tractable in many cases. This is the graphical modelling approach of *objective Bayesian networks*. This approach was originally applied to finite propositional inductive logic (Williamson 2005b, 2008; Haenni et al. 2011), but has been extended to first-order OBIL (Landes et al. 2024).

An objective Bayesian network (OBN) is a Bayesian network that represents the maximal entropy probability function $P^\dagger$. It consists of a directed acyclic graph whose vertices are the atomic sentences that feature in the support inference, together with the probability distribution of each atomic sentence conditional on each of its parents in the graph. From the OBN one can calculate the probability of the conclusion sentence to determine whether the entailment relationship holds. The distinctive feature of an OBN is that, unlike other kinds of Bayesian net, the construction of the directed acyclic graph is computationally trivial: there is no need for any computationally expensive probabilistic independence tests, for example.

Consider again the entailment relationship of the previous section. To construct an OBN we first construct an undirected graph $\mathcal{G}$ by taking atomic sentences that feature in the supports of the premisses as vertices and linking two atomic sentences if they occur together in one of the premiss supports. Thus, we take a_1, a_4 and a_{11} as vertices because they are the atomic sentences that feature in the premiss supports. We include an edge between a_1 and a_{11} because they both feature in $\check{\varphi}_3$, and an edge between a_1 and a_4 because they both feature in $\check{\varphi}_4$:

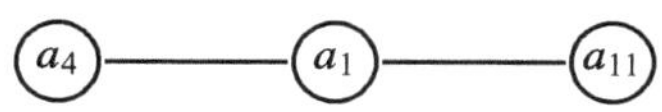

Maximal entropy probability functions render atomic propositions probabilistically independent in the absence of any evidence linking them. More precisely, one can prove that separation in $\mathcal{G}$ implies conditional probabilistic independence of the maximal entropy function $P^\dagger$ (Williamson 2002). In our example, $P^\dagger$ renders a_4 and a_{11} probabilistically independent conditional on a_1.

We next use a standard algorithm to transform $\mathcal{G}$ into a directed acyclic graph $\mathcal{H}$ that preserves as many of the conditional independencies of $\mathcal{G}$ as possible. For example, we can set $\mathcal{H}$ to be:

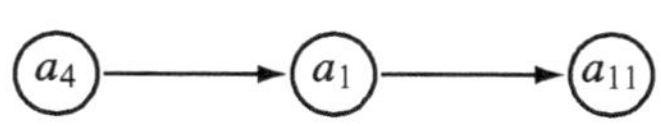

D-separation in $\mathcal{H}$ implies that $P^\dagger$ renders a_4 and a_{11} probabilistically independent conditional on a_1.[11]

[11] Subset Z D-separates subsets X from Y of nodes if each path between a node in X and a node in Y contains either (1) some node a_i in Z at which the arrows on the path meet head-to-tail ($\longrightarrow a_i \longrightarrow$) or tail-to-tail ($\longleftarrow a_i \longrightarrow$), or (2) some node a_j at which the arrows on the path meet head-to-head ($\longrightarrow a_j \longleftarrow$) and neither a_j nor any of its descendants are in Z. The key result is that if Z D-separates X from Y in $\mathcal{H}$ then the maximal entropy function renders X and Y probabilistically independent conditional on Z (Williamson 2005a, Theorem 5.3).

Finally, we determine the conditional probability distributions by finding the values of the following parameters that maximise entropy:

$$P(a_4),\quad P(a_1|a_4),\quad P(a_1|\neg a_4),\quad P(a_{11}|a_1),\quad P(a_{11}|\neg a_1).$$

A numerical optimisation yields:

$$P(a_4) = 0.9,\ P(a_1|a_4) = 1/3,\ P(a_1|\neg a_4) = 1/2,\ P(a_{11}|a_1) = 1,$$
$$P(a_{11}|\neg a_1) = 1/2.$$

Standard Bayesian network inference algorithms can then be used to determine the probability of a conclusion sentence of interest. For example,

$$\exists x Vt_1x,\quad (Ut_2 \vee \forall x Rx)^{0.9},\quad Ut_1 \rightarrow Vt_1t_3,\quad (Ut_1 \vee (\exists x Vxt_3 \rightarrow Ut_2))^{[0.95,1]}$$
$$\mid\!\approx (\neg(Ut_1 \vee Ut_3) \wedge \exists x Vxx \wedge Vt_1t_3)^{0.1625}$$

To see this, note that the support of the conclusion sentence is $\neg a_1 \wedge \neg a_7 \wedge a_{11}$ and that a_7 is not mentioned by any of the premisses so $P^\dagger$ renders a_7 probabilistically independent of a_1 and a_{11} and $P^\dagger(a_7) = 1/2$. Hence, $P^\dagger(\neg a_1 \wedge \neg a_7 \wedge a_{11}) = 1/2 \times P(a_{11}|\neg a_1)(P(\neg a_1|a_4)P(a_4) + P(\neg a_1|\neg a_4)P(\neg a_4)) = 1/2 \times 1/2\,(2/3 \times 9/10 + 1/2 \times 1/10) = 0.1625$.

A key advantage of OBNs over the truth table method is a reduction in the number of parameters required to specify the maximal entropy function $P^\dagger$. In this example, the support problem involves three atomic propositions and the truth table requires 8 parameters while the OBN requires only 5. Typically, this reduction in the number of parameters becomes very marked as the number of atomic propositions in the premisses increases.

3.8 Logical Properties

Here we briefly survey some of the logical properties of OBIL.

Some logical properties are common to all standard probabilistic logics (Landes 2023; Landes et al. 2024). For example:

Preservation of Deductive Entailment. If $\varphi \models \psi$ then $\varphi \mid\!\approx \psi$.
Reflexivity. $\varphi^X \mid\!\approx \varphi^X$.

Landes et al. (2023, Corollary 24) prove the following basic fact:

Theorem 3.11 (Zero-One Law) *Every constant-free sentence is either an inductive contradiction or an inductive tautology.*

Moreover, Landes et al. (2024) show the following:

Preservation of Inductive Tautologies (PIT). If $\mathrel{|\!\approx^{\circ}} \psi$ and $\varphi_1^{X_1}, \ldots, \varphi_k^{X_k}$ have satisfiable support then $\varphi_1^{X_1}, \ldots, \varphi_k^{X_k} \mathrel{|\!\approx^{\circ}} \psi$.

Note that PIT implies that inductive contradictions are also preserved by premisses with satisfiable support. Hence, by the zero-one law,

Corollary 3.1 (Extended Zero-One Law) *Every constant-free sentence is given either probability 1 or 0 by premisses with satisfiable support.*

Landes (2023, §3) derives several logical properties, including:[12]

Modus Ponens. If $\varphi^X \mathrel{|\!\approx^{\circ}} \psi^Y$ and $\mathrel{|\!\approx^{\circ}} \varphi^X$ then $\mathrel{|\!\approx^{\circ}} \psi^Y$.
Modus Tollens. If $\varphi \mathrel{|\!\approx^{\circ}} \psi$ and $\mathrel{|\!\approx^{\circ}} \neg\psi$ then $\mathrel{|\!\approx^{\circ}} \neg\varphi$.
Implication. If $\varphi \mathrel{|\!\approx^{\circ}} \psi$ then $\mathrel{|\!\approx^{\circ}} \varphi \rightarrow \psi$.
Weak Cautious Monotonicity. If $\varphi^X \mathrel{|\!\approx^{\circ}} \psi^Y$, $\varphi^X \mathrel{|\!\approx^{\circ}} \theta^Z$ and $\not\mathrel{|\!\approx^{\circ}} \neg\varphi$ then $\varphi^X, \psi^Y \mathrel{|\!\approx^{\circ}} \theta^Z$, as long as $X \neq \{0\}$.
Cautious Cut. If $\varphi^X \mathrel{|\!\approx^{\circ}} \psi^Y$ and $\varphi^X, \theta^Z \mathrel{|\!\approx^{\circ}} \psi^Y$ then $\varphi^X \mathrel{|\!\approx^{\circ}} \theta^Z$.
Very Cautious Or. If $\varphi \mathrel{|\!\approx^{\circ}} \psi, \theta \mathrel{|\!\approx^{\circ}} \psi$, $\not\mathrel{|\!\approx^{\circ}} \neg\varphi$ and $\not\mathrel{|\!\approx^{\circ}} \neg\theta$ then $\varphi \vee \theta \mathrel{|\!\approx^{\circ}} \psi$.

However, versions of some standard rules of nonmonotonic logic do not hold in general in OBIL (Landes 2023, §§4–5). For example,

Rational Monotonicity. If $\varphi^X \mathrel{|\!\approx} \psi^Y$ and $\varphi^X \not\mathrel{|\!\approx} \neg\theta$ then $\varphi^X, \theta \mathrel{|\!\approx} \psi^Y$.
Cautious Monotonicity. If $\varphi^X \mathrel{|\!\approx} \psi^Y$ and $\varphi^X \mathrel{|\!\approx} \theta^Z$ then $\varphi^X, \psi^Y \mathrel{|\!\approx} \theta^Z$.
And. If $\varphi^X \mathrel{|\!\approx} \psi^Y$ and $\varphi^X \mathrel{|\!\approx} \theta^Y$ then $\varphi^X \mathrel{|\!\approx} (\psi \wedge \theta)^Y$.
Transitivity. If $\varphi^X \mathrel{|\!\approx} \psi^Y$ and $\psi^Y \mathrel{|\!\approx} \theta^Z$ then $\varphi^X \mathrel{|\!\approx} \theta^Z$.

Interesting open questions remain. For example, is there a sound and complete set of rules for OBIL?

3.9 Language Invariance

An inductive logic is *language invariant* if, whenever $\varphi_1, \ldots, \varphi_k, \psi$ are sentences that can be formulated in both $\mathcal{L}^1$ and $\mathcal{L}^2$,

$$\varphi_1^{X_1}, \ldots, \varphi_k^{X_k} \mathrel{|\!\approx^1} \psi^Y \text{ if and only if } \varphi_1^{X_1}, \ldots, \varphi_k^{X_k} \mathrel{|\!\approx^2} \psi^Y,$$

where $\mathrel{|\!\approx^1}, \mathrel{|\!\approx^2}$ are the entailment relations defined on $\mathcal{L}^1, \mathcal{L}^2$ respectively. An inductive logic may be language invariant on a specific class of entailment relationships, even if it is not language invariant simpliciter.

[12] Note that in deriving these results, Landes assumed that any conclusion follows from unsatisfiable premisses ('explosion').

What is known about language invariance is generally positive:

Theorem 3.12 *OBIL is language invariant:*

- *for languages that differ only with respect to predicate symbols, on the class of finitely generated entailment relationships.* (*Williamson 2017, Theorem 5.9.*)
- *for languages whose constant symbols are permuted, as long as the permutation* σ *is such that* $\{t_1, \ldots, t_n\} = \{t_{\sigma(1)}, \ldots, t_{\sigma(n)}\}$ *for sufficiently large* n. (*Williamson 2010, Proposition 5.10.*)
- *for languages whose constant symbols are permuted, as long as the permutation* σ *preserves the set* $\mathbb{E} = [\varphi_1^{X_1}, \ldots, \varphi_k^{X_k}]$. (*Landes et al. 2023, Proposition 49.*)

A key question for further research concerns the extent to which these results can be generalised.

3.10 The Entropy-Limit Conjecture

The maximal entropy principle is an extension of the maximum entropy principle to the infinite predicate language $\mathcal{L}$. It is not the only such extension, however: Barnett and Paris (2008) put forward another approach to generalising the maximum entropy principle from a finite domain to an infinite predicate language $\mathcal{L}$. The question thus arises as to the relationship between the two approaches.

The core idea of Barnett and Paris (2008) is to consider a finite sublanguage $\mathcal{L}_n$ in which the premisses are expressible, and to re-interpret the premisses as saying something about a finite domain of $\mathcal{L}_n$, rather than as propositions about the infinite domain of $\mathcal{L}$. Thus $\forall x\theta(x)$ is re-expressed as $\theta(t_1) \wedge \cdots \wedge \theta(t_{r_n})$ and $\exists x\theta(x)$ is re-expressed as $\theta(t_1) \vee \cdots \vee \theta(t_{r_n})$. Then one can find the function $P^{(n)}$ that maximises n-entropy subject to the constraints imposed by the re-expressed premisses. One can consider the behaviour of $P^{(n)}$ as n increases by taking the limit $P^{(\infty)}(\psi) \stackrel{\text{df}}{=} \lim_{n\to\infty} P^{(n)}(\psi)$, where it exists. Finally, one can provide *entropy-limit* semantics for inductive logic on $\mathcal{L}$ itself: $\varphi_1^{X_1}, \ldots, \varphi_k^{X_k} \mathrel{|\!\!\!\approx}^{\infty} \psi^Y$ if and only if $P^{(\infty)}(\psi) \in Y$.

As it stands, this inductive logic is only partially defined, as the limit function $P^{(\infty)}$ does not always exist—for example, in certain cases where there are categorical Σ_2 or Π_2 premisses. However, Williamson (2017, p. 191) put forward the following conjecture:

Entropy-limit conjecture. Entropy-limit entailment $\mathrel{|\!\!\!\approx}^{\infty}$ agrees with maximal-entropy entailment $\mathrel{|\!\!\!\approx}^{\circ}$ wherever the limit $P^{(\infty)}$ exists and is in $\mathbb{E}$.

This conjecture has been explored in a number of cases and found to hold in each:

Theorem 3.13 *The entropy-limit conjecture has been verified to hold in the following special cases:*

- *Inferences with categorical monadic premisses.*

(Barnett and Paris 2008; Rafiee Rad 2009, Theorem 29; Rafiee Rad 2021)

- *Inferences with categorical Σ_1 premisses.* *(Rafiee Rad 2018)*
- *Inferences with categorical Π_1 premisses.* *(Landes et al. 2021)*
- *Inferences from a premiss of the form φ^c, where the entropy-limit conjecture holds for both φ and $\neg\varphi$.* *(Landes et al. 2021)*
- *Certain inferences in which $P^{(\infty)}$ is a limit in entropy of the $P^{(n)}$.* *(Landes et al. 2021)*

If the entropy-limit conjecture were to hold generally, this would support the claim that there is a canonical inductive logic and that OBIL captures that logic. Thus, the status of the entropy-limit conjecture is one of the central open questions in this field.

3.11 Conclusions

As we have seen, the maximal entropy principle generalises Jaynes' maximum entropy principle and underpins an inductive logic defined on a first-order predicate language. This logic, objective Bayesian inductive logic, suffers neither from the underdetermination of the Carnapian and precise Bayesian approaches nor from the weak inferences of an imprecise Bayesian approach. A large class of entailment relationships (including all those that satisfy closure in entropy) induce a unique maximal entropy function, and a large subclass of those entailment relationships (the class of those whose premisses have satisfiable support) is known to be decidable. These entailment relationships can be verified by means of truth tables or objective Bayesian nets. OBIL has been found to be language independent in key cases, and to agree with the alternative approach of Barnett and Paris (2008) in key cases. Thus OBIL is now a mature and promising theory.

But there remain many open questions, not just concerning language independence and the entropy-limit conjecture, but also, for example, the question of whether OBIL can be characterised by logical rules, the extent to which limits in entropy exist, and the behaviour of OBIL in cases where $\mathbb{E}$ is not closed in entropy. Thus, there are many opportunities to advance the theory and develop our understanding of OBIL. Enough has been done to demonstrate the potential of OBIL, but not so much as to rob us of exciting new results.

Acknowledgments Most of the results discussed here are the product of joint research with Jürgen Landes and Soroush Rafiee Rad. This research was supported by Leverhulme Trust grant RPG-2022-336.

References

Barnett, O., and J. Paris. 2008. Maximum entropy Inference with Quantified Knowledge. *Logic Journal of the IGPL* 16 (1): 85–98.
Boyd, S., and L. Vandenberghe. 2004. *Convex Optimization*. Cambridge: Cambridge University Press.
Carnap, R. 1952. *The Continuum of Inductive Methods*. Chicago: University of Chicago Press.
de Finetti, B. 1937. Foresight. Its Logical Laws, Its Subjective Sources. In *Studies in Subjective Probability*, ed. H. E. Kyburg and H. E. Smokler, 2nd (1980) ed., 53–118. Huntington: Robert E. Krieger Publishing Company.
Haenni, R., J.-W. Romeijn, G. Wheeler, and J. Williamson. 2011. *Probabilistic Logics and Probabilistic Networks*. Synthese Library. Dordrecht: Springer.
Howson, C. 2000. *Hume's Problem: Induction and the Justification of Belief*. Oxford: Clarendon Press.
Howson, C., and P. Urbach. 1989. *Scientific Reasoning: The Bayesian Approach*. 3rd (2006) ed. Chicago: Open Court.
Jaynes, E.T. 1957. Information Theory and Statistical Mechanics. *The Physical Review* 106 (4): 620–630.
Jaynes, E.T. 1979. Where Do We Stand on Maximum Entropy? In *The Maximum Entropy Formalism*, ed. R. D. Levine, and M. Tribus, 15–118. Cambridge: MIT Press.
Jaynes, E.T. 2003. *Probability Theory: The Logic of Science*. Cambridge: Cambridge University Press.
Jeffrey, R. 2004. *Subjective Probability: The Real Thing*. Cambridge: Cambridge University Press.
Keynes, J.M. 1921. *A Treatise on Probability*. 1973 ed. London: Macmillan.
Landes, J. 2023. Rules of Proof for Maximal Entropy Inference. *International Journal of Approximate Reasoning* 153: 144–171.
Landes, J., and J. Williamson. 2013. Objective Bayesianism and the Maximum Entropy Principle. *Entropy*, 15 (9): 3528–3591.
Landes, J., and J. Williamson. 2015. Justifying Objective Bayesianism on Predicate Languages. *Entropy* 17 (4): 2459–2543.
Landes, J., S. Rafiee Rad, and J. Williamson. 2021. Towards the Entropy-Limit Conjecture. *Annals of Pure and Applied Logic* 172 (2): 102870.
Landes, J., S. Rafiee Rad, and J. Williamson. 2023. Determining Maximal Entropy Functions for Objective Bayesian Inductive Logic. *Journal of Philosophical Logic* 52: 555–608.
Landes, J., S.Rafiee Rad, and J. Williamson. 2024. A Decidable Class of Inferences in First-Order Objective Bayesian Inductive Logic. *Under review*.
Nilsson, N.J. 1986. Probabilistic Logic. *Artificial Intelligence* 28: 71–87.
Paris, J.B. 1994. *The Uncertain Reasoner's Companion*. Cambridge: Cambridge University Press.
Rafiee Rad, S. 2009. *Inference Processes for Probabilistic First Order Languages*. Ph.D. Thesis, Department of Mathematics, University of Manchester.
Rafiee Rad, S. 2018. Maximum Entropy Models for Σ_1 Sentences. *Journal of Applied Logics - IfCoLoG Journal of Logics and Their Applications* 5 (1): 287–300.
Rafiee Rad, S. 2021. Probabilistic Characterisation of Models of First-Order Theories. *Annals of Pure and Applied Logic* 172 (1): 102875.
Ramsey, F.P. 1926. Truth and Probability. In *Studies in Subjective Probability*, ed. H. E. Kyburg, and H. E. Smokler, 2nd (1980) ed., 23–52. Huntington/New York: Robert E. Krieger Publishing Company.
Rosenkrantz, R.D. 1977. *Inference, Method and Decision: Towards a Bayesian Philosophy of Science*. Dordrecht: Reidel.
Seidenfeld, T. 1986. Entropy and Uncertainty. *Philosophy of Science* 53 (4): 467–491.
Shannon, C. 1948. A Mathematical Theory of Communication. *The Bell System Technical Journal* 27: 379–423, 623–656.
Topsøe, F. 1979. Information Theoretical Optimization Techniques. *Kybernetika* 15: 1–27.

Williamson, J. 2002. Maximising Entropy Efficiently. *Electronic Transactions in Artificial Intelligence Journal* 6.
Williamson, J. 2005a. *Bayesian Nets and Causality: Philosophical and Computational Foundations*. Oxford: Oxford University Press.
Williamson, J. 2005b. Objective Bayesian Nets. In *We Will Show Them! Essays in Honour of Dov Gabbay*, ed. S. Artemov, H. Barringer, A.S. d'Avila Garcez, L.C. Lamb, and J. Woods, vol. 2, 713–730. London: College Publications.
Williamson, J. 2008. Objective Bayesian Probabilistic Logic. *Journal of Algorithms in Cognition, Informatics and Logic* 63: 167–183.
Williamson, J. 2010. *In Defence of Objective Bayesianism*. Oxford: Oxford University Press.
Williamson, J. 2017. *Lectures on Inductive Logic*. Oxford: Oxford University Press.
Williamson, J. 2018. Justifying the Principle of Indifference. *European Journal for the Philosophy of Science* 8: 559–586.
Williamson, J. 2023. Bayesianism from a Philosophical Perspective and Its Application to Medicine. *International Journal of Biostatistics* 19 (2): 295–307.

Chapter 4
Probability Logic and Statistical Relational Artificial Intelligence

Felix Weitkämper

Abstract After decades of study in the philosophy community, probability logic established itself as a knowledge representation formalism for quantifying uncertainty in artificial intelligence in the late 1980s. Since then, however, statistical relational artificial intelligence emerged as a new paradigm for combining first-order reasoning with probabilistic uncertainty, harnessing the rapid advances made in the field of probabilistic graphical models. In this contribution, we suggest how first-order logics of probability, and the taxonomy developed for them around 30 years ago, can contribute to a deeper understanding of statistical relational frameworks, enhance the expressivity of their models and formulate queries of practical significance and theoretical interest.

Keywords Statistical relational AI · Machine learning · Probabilistic logic · Probabilistic logic programming · Bayesian networks

4.1 Introduction

For most of its history, artificial intelligence was intimately connected with the art of automated reasoning. In a classical system of symbolic artificial intelligence, knowledge is encoded by experts as facts and rules in a suitable formal language. Using those, new knowledge can be derived by reasoning. Facts and rules could be highly sophisticated, involving intricate relations between different entities. Paradigmatic for this approach are expert systems and other artificial intelligence applications

---

Some of the content covered in this chapter has previously been discussed in my columns in The Reasoner magazine, www.thereasoner.org

F. Weitkämper (✉)
Institut für Informatik, Ludwig-Maximilians-Universität München, München, Germany
e-mail: felix.weitkaemper@lmu.de

H. Hosni, J. Landes (eds.), *Perspectives on Logics for Data-driven Reasoning*,
Logic, Argumentation & Reasoning 35,
https://doi.org/10.1007/978-3-031-77892-6_4

based on *logic programming*, and in particular on the language Prolog and its dialects. While this leads to a highly expressive way of representing knowledge, a major limitation soon emerged: The world is indeed complex, but it is also uncertain, and general rules hardly hold without exception. Although probability theory and statistics are well-developed, time-honoured parts of science, incorporating probabilistic reasoning into artificial intelligence was long considered out of the question, since it would be impossible to specify all conceivable correlations and dependencies between factors. However, with the introduction of Bayesian networks in the late 1980s, transparent independence assumptions suddenly made this viable. Since then, statistical machine learning with probabilistic graphical models has developed into a large area of artificial intelligence. In the process, the emphasis shifted from reasoning about complex relationships to learning the dependencies between simple properties.

Independently of these efforts, dedicated logics have been developed for reasoning with and about probabilities. While the roots of these efforts go back to Carnap (1950) and the philosophical movement of inductive logic that he brought in his wake, the artificial intelligence community embraced first-order probabilistic logics around the late 1980s, among others due to the efforts of Bacchus (1990) and Halpern (1990). However, unlike statistical machine learning, probability logics have never broken through from knowledge representation to the mainstream of practical learning and reasoning. Instead, since the mid-1990s, a new field emerged of Statistical relational artificial intelligence (sometimes abbreviated StarAI), which aims to combine probabilistic reasoning with the expressivity of relational logic not from the viewpoint of probability logic, but by combining concepts from statistical machine learning with first-order logic and logic programming.

In this chapter, I will explore the relationship between probability logic and statistical relational artificial intelligence, and outline how I think classical probability logic can be made fruitful for the study and use of statistical relational approaches.

To give the discussion some substance, the following subsections introduce prototypical examples of statistical relational formalisms and first-order logics of probability. The following section then makes three concrete proposals as to how the classical field of probability logic can be made fruitful for statistical relational artificial intelligence, namely as a means to provide a unified semantics, as a component in specifying a statistical relational model, and as a query language.

4.1.1 Statistical Relational Artificial Intelligence

Combining logic programming and probabilistic graphical models can be tackled from either direction: One can expand logic programming to incorporate probabilities, or one can extend graphical models to incorporate relational information. As there are a plethora of different formalisms in either of the two categories, we restrict ourselves here to introducing one approach of each type. Even so, we refer to the literature for further details, and in particular we presuppose some intuitive understanding of Bayesian networks and logic programs. The concise introductory volume by Raedt et al. (2016) provides a more self-contained introduction to the field.

4.1.1.1 Probabilistic Logic Programming

We start with a discussion of arguably the most popular extension of logic programming with probabilities: *probabilistic logic programming under the distribution semantics*.

The key idea behind probabilistic logic programming is the *distribution semantics*, introduced independently by Poole (1993) and Sato (1995). The distribution semantics neatly divides a probabilistic logic program into a simple list of probabilistic facts and an ordinary logic program which takes those probabilistic facts as input to compute more complex predicates. More precisely, a probabilistic logic program consists of a list of assertions, each annotated with a probability between 0 and 1, and a logic program whose extensional vocabulary coincides with the signature of those facts. We illustrate this idea with a common toy example from the literature:

Example: Friends and Smokers

The probabilistic logic program *Friends and Smokers* consists of the probabilistic facts

```
0.2 :: befriends(X,Y).
0.5 :: influences(X,Y).
0.3 :: stress(X).
```

and the rules

```
friends(X,Y) :- befriends(X,Y).
friends(X,Y) :- befriends(Y,X).
smokes(X)   :- stress(X).
smokes(X)   :- friends(X,Y), smokes(Y), influences(Y,X).
```

The semantics of this program reads as follows: For every domain entity (referred to in the following as a *person*), there is a 30% chance that this person is stressed (that is, `stress(person)` is true). Similarly, for every pair of persons, there is a 20% chance that one person befriends the other, and a 50% chance that one person influences the other. All these random choices are made independently of one another. After these choices have been made, the rules of the program are brought to bear. In this case, the binary relation `friends` is evaluated as the symmetric closure of `befriends`, and the smokes relation is computed as follows: First, every stressed individual smokes. Then, `smokes` predicate is spread recursively along friends influencing each other. Once this process has been completed and no more smokers can be added this way, the program terminates.

On the first glance, this formalism seems very restrictive, since it confines probability entirely to independent choices on factual assertions. However, the ingenuity of the distribution semantics is that probabilistic rules can be expressed by simply adding additional auxiliary predicates, such as `influences` in the example below:

Example: Friends and Smokers revisited

The probabilistic rule

```
0.5 :: smokes(X) :- friends(X,Y), smokes(Y).
```

is equivalent to the clauses

```
smokes(X)  :- friends(X,Y), smokes(Y), influences(Y,X).
  0.5 :: influences(X,Y).
```

of the program in the first example.

In fact, it has been shown that if only ground (propositional) clauses are considered, acyclic probabilistic logic programs have the same expressiveness as Bayesian networks, while general probabilistic logic programs go beyond that by adding recursion (Riguzzi and Swift 2018).

However, the real strength of probabilistic logic programs (and statistical relational methods in general) lies in the ability to specify rules and facts with variables, and therefore to specify a particular probability distribution for any domain under consideration. From this viewpoint the probabilistic logic program associates with every domain of people a probability distribution over possible L-structures on this domain, where L contains all the predicates mentioned in the program. In this context L-structures are usually referred to as *possible worlds*. The following description of the semantics is not completely formal, since some assumptions on the probabilistic logic program have been omitted, but serves to give an idea:

The probability of a possible world is evaluated in two phases. First, one checks whether the *intensional* predicates, those occurring in the head of logical rules, are indeed the result of evaluating those rules on the *extensional* predicates, those that are associated with probabilistic facts. If not, this world cannot possibly occur and is assigned probability 0. If yes, then the probability of the world is determined simply by multiplying up the probabilities of all the probabilistic facts $p_R : R(\mathbf{x})$ true in the world with the product of $(1 - p_R)$ for all probabilistic facts that are *false* in that world.

This framework can be extended to consider as input not just plain domains but structures in an extensional vocabulary $E \subseteq L$; for instance, the network of friendships in the examples here could be taken as input rather than generated probabilistically. In that case, every input structure M gives rise under a probabilistic logic program to a probability distribution over all possible L-structures extending

M. For a more detailed contemporary treatment of probabilistic logic programming under the distribution semantics, see the comprehensive textbook by Riguzzi (2023).

The tasks for any probabilistic logic program system can be divided into two main headers: inference and learning.

Among inference tasks, the most prominent are marginal inference, where the probabilities of a certain property are computed, and maximum-a-posteriori (MAP) inference, where the most likely configuration to arise from a probabilistic logic program is computed. In the example above, a marginal query could compute the probability of a certain individual smoking. MAP inference could answer the question: What is the most likely smoking behaviour of this group of individuals given some observations?

Unfortunately, the naïve approach to inference, that is, substituting the domain individuals into the probabilistic logic program (grounding) and then performing inference on the ground program does not scale well to large datasets. Therefore, current research is much concerned with lifted inference, which is performed to varying extent on the level of the original probabilistic logic program (with variables) rather than on the large grounded program.

The problem of learning probabilistic logic programs from data is posed in different settings. In parameter learning, the rules of the program are fixed and an optimal set of probabilities for those rules are sought. In structure learning, the rules themselves are not given either and have to be found by the learning algorithm. Structure learning of probabilistic logic programs is generally considered a difficult problem, since it subsumes and significantly extends the problem of learning deterministic logic programs, known as Inductive Logic Programming. Indeed, early structure learning algorithms approached structure-learning in two parts, applying first an Inductive Logic Programming system to learn the rules and then a parameter learning algorithm for the probabilities. However, better results have recently been obtained by interleaving the two parts.

I would like to close this subsection by highlighting two particularly active and well-maintained systems, both of which have an easy-to-use online interface and tutorial.

ProbLog 2 (Fierens et al. 2015), an implementation of the ProbLog language maintained by the KU Leuven group available at https://dtai.cs.kuleuven.be/problog/, is a powerful Python-based system with many features. Its underlying language, ProbLog, has been designed particularly with usability and straightforward syntax in mind, and it has been used extensively in bioinformatics. A variety of academic and real-world applications are listed at https://dtai.cs.kuleuven.be/problog/applications.html.

cplint (Riguzzi 2023), a SWI-Prolog-based system supporting various languages and maintained by the University of Ferrara, is available at https://cplint.ml.unife.it/. It is particularly rich in configurations and supported algorithms, including state-of-the-art methods for structure learning and lifted inference (neither of which are supported by ProbLog 2 at present). On the other hand, it is not quite as straightforward to use as ProbLog 2.

4.1.1.2 First-Order Bayesian Network Specifications

As a contrasting example, we consider a framework based on lifting Bayesian networks. There are a large number of such frameworks in the literature, the arguably most-used and most general being Jaeger's (1997) *relational Bayesian networks*. For the sake of transparency and simplicity, we restrict ourselves here to a far more restrictive formalism that associates a single probability with each of a finite list of first-order sentences.

Definition 4.1 (First-Order Bayesian Networks) A *first-order Bayesian network* in a relational signature L consists of a directed acyclic graph whose nodes are the relation symbols from L, and where every node is associated with additional data as follows:

- If a node R has no parents, then it is annotated with a real number $p_R \in [0, 1]$.
- Otherwise, a node R is annotated with a tuple $\mathbf{x}$ of variables, whose length corresponds to the arity of R, a finite set of first-order formulas $\varphi_1, \ldots, \varphi_n$ with the properties below, and for every such formula a real number $p_{R,i} \in [0.1]$:

1. The free variables of every φ_i are exactly $\mathbf{x}$.
2. Only relation symbols among the parents of R occur in a φ_i.
3. For any $1 \leq i < j \leq n$, $\varphi_i \wedge \varphi_j$ is a contradiction.
4. The formula $\varphi_1 \vee \cdots \vee \varphi_n$ is a tautology.

We illustrate the semantics of such a network with the following toy model, where a house with neighbours is potentially affected by a burglary, and the police may be called.

Example: Burglaries and neighbours
Consider a signature of two unary relation symbols, Burglary and Alarm, and the binary relation symbol Neighbour. Then consider the DAG with only two edges, one from Burglary into Alarm and one from Neighbour into Alarm. Let Burglary have the annotation 0.05 and Neighbour the annotation 0.1. Alarm is annotated with the variable x and the formulas

$$\text{Neighbour}(x, h) \wedge \text{Burglary}(h)$$

and

$$\neg\left(\text{Neighbour}(x, h) \wedge \text{Burglary}(h)\right).$$

The first formula is associated with a value of 0.6, while the second formula is associated with a value of 0.01.

(continued)

The semantics of this network is that at every house, there is a chance of 0.05 that there is a burglary, and every pair of houses has a 0.1 chance of being neighboured. All those events are considered to be independent. Furthermore, if someone witnesses a burglary in a neighbouring house (that is, the first formula is satisfied), there is a 0.6 chance that this neighbour raises the alarm. If not, the probability that someone raises the alarm without having witnessed a burglary in a neighbouring house is 0.01.

Just as for probabilistic logic programs, external predicates given by data can be accommodated simply by omitting the probability annotation on (some) root predicates.

Relational Bayesian networks and similar formalisms usually allow for defining first-order formulas with free variables and then use some *aggregation function* to map the numbers of true and false groundings of those formulas to probabilities. This goes beyond the expressivity of the probability logics we discuss in this article, but can be accommodated by shifting to a framework of *continuous-valued* logic (Koponen and Weitkämper 2023a,b).

4.1.2 Probability Logic

To understand the relationship between StarAI and first-order logics of probability, one must first dive into a fundamental philosophical question: What *is* probability? One answer that proved particularly influential in the development of logic was given by Rudolf Carnap, who postulated in his 1945 paper *The two concepts of probability* that in fact the term is used to denote two different concepts.

4.1.2.1 Types of Probability Logic

One, which he terms probability$_1$, refers to the degree of confirmation of statements; a typical question of probability$_1$ would be, "Given that one has observed three black sheep, what is the probability that all sheep are black?" This question seemed to Carnap an intrinsically *logical* question, if a very hard on at first. There is no evidence needed to answer it, or rather, the evidence is part of the question itself. The other concept, probability$_2$, refers to the relative frequency of statements; a typical question would be "What is the probability of a given sheep to be black?" The intended answer to this question is *empirical*: The answer is precisely the relative frequency of black sheep, which happens to be less than 25% according to the latest scholarly sources.

After a long time of reasonably active study in the philosophy community, and particularly in a subfield that came to be known as pure inductive logic, computer scientist Joe Halpern introduced these concepts to artificial intelligence in his seminal 1990 paper "An analysis of first-order logics of probability". Halpern called logics that address questions about probability$_2$ *Type I logics*, and those that address questions about probability$_1$ he called *Type II logics*, unfortunately exactly inverting Carnap's numbering.

The relevance of this type distinction to probability logic stems from the different semantics of probabilistic statements entailed by the two viewpoints. A logic designed for answering probability$_2$-statements is evaluated over a single structure, much like a sentence of first-order logic would be, but its domain is additionally equipped with a probability measure over its elements. Thus, one can answer the example question above by evaluating the probability of the event $\{x \in \omega \mid \omega \models \text{Black}(x)\}$ on the probability space of domain elements.

A logic designed for answering probability$_1$-statements has a very different semantics: It is evaluated not over a probability space of domain elements, but over a probability space of possible worlds. In other words, rather than a measure on the domain of a single structure, a potential model of a probability$_2$ sentence is a probability measure P on a space of L-structures (usually those of all L-structures with a given domain). In that way, the answer to the sample question above would be the conditional probability

$$P\left(\omega \models \forall_x \text{Black}(x) \mid \omega \models \text{Black}(a) \wedge \text{Black}(b) \wedge \text{Black}(c)\right).$$

Halpern also found that these notions of probability could easily be combined, leading to a *Type III logic* whose models are probability distributions over worlds, each of whose domains is also equipped with a probability measure.

4.1.2.2 First-Order Logics of Probability

We present here two simple, exemplary logics for Type I and Type II probability inspired by those of Halpern (1990). As independence statements are particularly important for encoding statistical relational frameworks, we also introduce an extension of Type II logic allowing for independence statements to be expressed directly.

As in the preceding section, we assume a basic familiarity with first-order logic. To avoid overfreighting our notation, we also make use of the convenient shorthand $\varphi(\mathbf{x})$ to emphasis that all free variables occurring in φ are from $\mathbf{x}$.

For a Type I logic, consider this variant of relational first-order logic: As usual, variables and constants are terms, and an atomic formula is an n-ary relation symbol R applied to an n-tuple of terms. General formulas are obtained from atomic formulas by negation ($\neg$) and conjunction ($\wedge$), but also by applying *probabilistic quantifiers* of the form $\mathbb{P}_x(\varphi) \leq p$ for a real number $p \in [0, 1]$. Probabilistic quantifiers bind the variable x as usual.

As explained above, the semantics of this Type I logic with signature L is given by L-structures that are also endowed with a probability measure μ on their domain, and formulas are evaluated against such a structure in the usual way, where $\mathbb{P}_x(\varphi) \leq p$ holds if "The μ-measure of the set of domain elements x for which φ holds is at most p".

Type I logics used in practice are often extensions of this simple system, which also allow for taking the probability of tuples of variables rather than merely single variables and directly expressing conditional probabilities. Such a more elaborate Type I logic is introduced for instance by Koponen (2020) for use in lifted Bayesian networks.

Let us now consider a simple example of a Type II logic. Terms and first-order atomic formulas are as in relational first-order logic. First-order formulas are constructed from those by applying the full range of first-order connectives and quantifiers. An atomic *probability formula* is of the form $\mathbb{P}(\varphi) \leq p$ for a real number $p \in [0, 1]$ and a first-order formula φ. The free variables of an atomic probability formula are exactly those of φ. Then a Type II formula is constructed from atomic probability formulas by again applying first-order quantifiers.

Unlike its Type I counterpart, formulas of Type II probability logic are evaluated against a domain D and a *family* Ω of L-structures with domain D, which is equipped with a probability measure μ. Note that while in Type I probability logic the measure was taken over the domain, it is now taken over a set of possible structures on the domain.

This is also reflected in the semantics of the Type II logic. For every structure $\omega \in \Omega$, first-order formulas are evaluated as usual. Then for every formula φ and interpretation ι of the free variables in φ, $\mathbb{P}(\varphi) \leq p$ holds if the measure of the set of worlds in which φ holds under the interpretation ι does not exceed p. Finally, the truth values of the overall formulas are derived from this by applying the usual meaning of the first-order connectives to the probability atoms.

Although this "official" definition only uses one-directional inequalities, one can define $\mathbb{P}(\varphi) \geq p$ as a shorthand for $\mathbb{P}(\neg\varphi) \leq 1 - p$, and $\mathbb{P}(\varphi) = p$ as a shorthand for $\mathbb{P}(\varphi) \leq p \wedge \mathbb{P}(\varphi) \geq p$

For the purposes of statistical relational artificial intelligence, probabilistic independence plays a crucial role. Thus, we extend the simple Type II logic above with an additional probability quantifier, $\perp$, used as follows: If $\varphi_1, \ldots, \varphi_n$ are first-order formulas, then $\perp_{\mathbf{x}} (\varphi_1, \ldots, \varphi_n)$ is a probability atom, whose free variables $\mathbf{y}$ are all those variables occurring in a φ_i but not in $\mathbf{x}$. Given an interpretation ι of its free variables, $\perp_{\mathbf{x}} (\varphi_1, \ldots, \varphi_n)$ is true if the set

$$\{\varphi_i(\mathbf{a}, \iota(\mathbf{y})) \mid \mathbf{a} \in D, 1 \leq i \leq n\}$$

is a mutually independent set of events, where $\varphi_i(\mathbf{a}, \iota(\mathbf{y}))$ is the event "φ_i is true in ω under the extension of ι which maps $\mathbf{x}$ to $\mathbf{a}$". This semantics was inspired by the propositional probability logic with conditional independence presented by Ivanovska and Giese (2010), and just as their version is designed specifically to express the independencies enforced by Bayesian networks, ours is designed to express the independencies inherent in a set of probabilistic facts.

More expressive versions of the Type II logic above, such as the one originally proposed by Halpern (1990), also incorporate polynomial dependencies between the probabilities, which render the independence operator definable within the more general language. Apart from the notational convenience afforded by using the $\perp$ operator directly, avoiding direct polynomial comparisons between probabilities of formulas could also reduce the complexity of inference (Ivanovska and Giese 2010).

We refer to extensive discussions by Keisler (1985), Abadi and Halpern (1994), Perovic et al. (2008), and Ognjanovic et al. (2016) for more details on the complexity of inference and model checking and for discussions of (generally infinitary) proof calculi for Type I and Type II logics of probability.

4.2 Probability Logic and Statistical Relational Artificial Intelligence

Comparing the analysis of different probability types to what we know about statistical relational artificial intelligence, one perspective on its relationship to probability logic emerges. For any specified domain, a statistical relational specification defines a probability distribution over the possible worlds with that domain. In this sense, one could argue that statistical relational languages *are* Type II probability logics. They are declarative, syntactical specifications whose semantics is given by a probability measure over possible worlds.

However, there are two key differences that I believe distinguish statistical relational approaches from traditional probability logic. The first is that they arrive at a probabilistic semantics by augmenting a non-probabilistic logic with an essentially separate and non-logical or logically trivial probabilistic formalism, either probabilsitic graphical models or, in the case of probabilistic logic programs, simple probabilistic facts. There is another key difference to traditional probability logics, though: While in traditional probability logic, a sentence can hold either zero, one or multiple probability distributions, a statistical relational framework defines exactly one single probability measure for any domain. This reminds one strongly of the traditional relationship between classical logic and logic programming, where a sentence of first-order logic can have any number of models, while the meaning of a logic program is defined as its one unique minimal model.

4.2.1 Completion Semantics for StarAI

A hallmark event in the history of logic programming was Keith Clark's (1978) paper on the *completion semantics*, in which he identified a non-recursive logic program with the *completion* of its associated logical rules (Clark 1978). Consider the following example:

Example: Clark Completion

The logic program

```
p(X) :- q(X).
p(X) :- q(Y), r(X,Y).
q(1).
r(0,1).
```

naturally expresses the logical statements

$$\forall_X \, (p(X) \leftarrow q(X)),$$

$$\forall_X \, (p(X) \leftarrow \exists_Y \, (q(Y) \wedge r(X, Y)))$$

as well as asserting $q(1)$ and $r(0, 1)$.

However, the logical theory defined by those statements has more than one model, even if the domain is fixed as comprising only the two elements 0 and 1 (the *Herbrand base* of this logic program). For instance, $q(0)$ could be true, or not.

The completion of the program adds the following sentences, so that the resulting larger theory has exactly one model with any given domain:

$$\forall_X \, (p(X) \rightarrow (q(X) \vee \exists_Y \, (q(Y) \wedge r(X, Y)))),$$

$$q(X) \rightarrow X = 1,$$

$$r(X, Y) \rightarrow X = 0 \wedge Y = 1.$$

The idea behind Clark's completion is to augment the first-order rules naturally encoded by a logic program, which will generally have more than one model, with additional rules that pin down exactly the one model that is the intended semantics of the program.

If the logic program uses recursion, then the completed theory as introduced here can also have more than one model and thus itself fail to pin down the meaning of the logic program. In *stratified programs*, when negation is not intertwined with recursion, this can be remedied by moving beyond first-order logic and introducing an explicit fixed-point operator to the language. Such a generalised completion semantics was first introduced in database theory by Abiteboul and Vianu (1989, 1991), and explicitly proposed as a completion semantics for logic programs by Warren and Denecker (2023). Ebbinghaus and Flum (1999, Chapters 8 and 9) give a detailed presentation of the logical issues surrounding recursive logic programs.

To extend this idea to probabilistic logic programming, we can take the Clark completion semantics of the associated logic program as a basis. The probabilistic facts, of the form $\alpha :: R(\mathbf{x})$, naturally encode the Type II statement that for all $\mathbf{x}$, the probability of $R(\mathbf{x})$ is α. However, just as in classical logic programming, the naturally encoded statements are insufficient to pin down an exact model. This happens because they do not rule out arbitrary probabilistic dependencies between the atomic events associated with the probabilistic facts. For instance, it could be that all $R(\mathbf{x})$ are always simultaneously true or false, with both cases having probability α and $1 - \alpha$ respectively, or they could all be mutually independent events. In the probabilistic independence logic that we sketched out in Sect. 4.1.2.2 above, we can express that indeed the set of all atomic events associated with a probabilistic fact is an independent set with a statement of the form

$$\perp_{\mathbf{x_1},\ldots,\mathbf{x_n}} \{a_1(\mathbf{x_1}), \ldots, a_n(\mathbf{x_n})\},$$

where the a_i range over all probabilistic facts in the model and the $\mathbf{x_1}, \ldots \mathbf{x_n}$ are arbitrary tuples of variables whose lengths match the arities of the associated probabilistic facts. The semantics of this statement is that the set of all groundings of probabilistic facts is mutually independent, which precisely pins down the probability distribution defined by the probabilistic facts. Together with the completion of the associated logic program, this sentence thus uniquely determines the probability distribution induced by the probabilistic logic program.

If we interpret this independence statement as "there are no mutual dependencies beyond those implies by the logic program", then we see the direct parallel to Clark's completion for ordinary logic programs, which encoded "There are no cases in which the head atom is true beyond those implies by the logic program".

When negation and recursion occur together, there are several possible ways to proceed. The most widely used alternatives for this case are the well-founded semantics and the stable model semantics (Truszczynski 2018). However, neither of those guarantee a unique model model in a 2-valued logic. In answer set programming, for instance, which is based on the stable model semantics, a general recursive logic programming is associated with any number of *answer sets* or stable models. While not originally defined as a completion semantics, the stable models of a propositional logic program can be characterised as models satisfying Clark's completion and additionally a set of propositional *loop* formulas that exclude non-well-founded computations (Lin and Zhao 2004).

When defining a probabilistic extension of answer set programming, one has to decide how to deal with this multiplicity of models for a single program. To my knowledge the first semantics proposed for probabilistic answer set programming was P-Log, which divides the probability mass equally between the stable models corresponding to a single answer set program. In that way, we return to a situation in which there is a single probability distribution (Baral et al. 2009). More recently, a different approach has gained traction, the *credal semantics*. In the credal semantics, a query returns two probabilities, an upper bound and a lower bound. For the upper bound, a configuration of random facts contributes to the query probability if the query is true in *any* of its associated answer sets; for the lower bound, a configuration of random facts contributes to the query probability if the query is true in *all* of its associated answer sets (Cozman and Mauá 2020).

Probability logic could allow for a new completion semantics for propositional probabilistic answer set programs under the credal semantics, where Clark's completion and the loop formulas are simply augmented with the independence of random facts as described above. Then, the upper and lower bound could be computed as the largest probability interval implied by the theory of probabilistic logic.

Ivanovska and Giese (2010) provide a semantics for propositional Bayesian networks as conditional probability statements augmented by independence statements in a probabilistic logic, which is very similar in spirit to the completion semantics for probabilistic logic programs suggested above. In principle, their axiomatisation can clearly be extended to the first-order Bayesian networks as we have defined them here. However, since their axiomatisation refers to conditional probabilities and requires statements regarding pairwise independence rather than merely the stronger fully independent sets expressible in our model Type II logic, one would have to use a more expressive language than the one we introduced above.

4.2.2 Type I Logics for StarAI

There seems to be significant untapped potential in utilising Type I probability logics in the defining formulas of statistical relational formalisms that are currently expressed in ordinary first-order logic. This would lift the expressivity of such frameworks from a Type II to a Type III probability logic, with all the same benefits as in the traditional setting of probability logic. While Azzolini et al. (2022) at the University of Ferrara have worked on integrating such statements into probabilistic logic programming, Vera Koponen from the University of Uppsala and I (Koponen 2020; Koponen and Weitkämper 2023a,b; Weitkämper 2022) have made some attempts in this direction in the context of frameworks based on Bayesian networks.

To get an idea of the expressive potential of Lifted Bayesian networks based on Conditional Probability Logic, the Type I logic used by Koponen (2020), consider the following example (Weitkämper 2022):

Example: Conditional Probability Logic
As an example combining several general features of statistical relational modelling with CPL, consider a policy where schools and workplaces are shut whenever there is either a diagnosed positive case in that school or workplace or when 1% of the population are diagnosed with the disease. Additionally, there is a higher chance of contact between two people if they attend the

(continued)

same open school or workplace. This can all be expressed in a lifted Bayesian network over a two-sorted signature as follows:

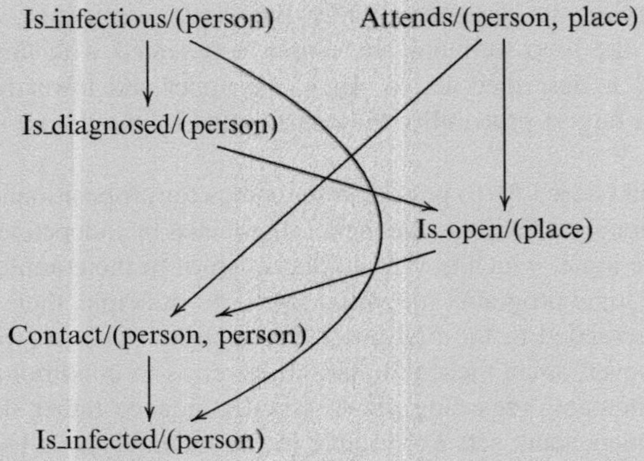

Assume that "Attends" is given by supplied data. Depending on whether this network is just one component in an iteration, where "Is_infectious" depends on the "Is_infected" of a previous time step, or stands alone, Is_infectious might be given by data. Alternatively, it might be stochastically modelled, with a certain fixed probability. Then the conditional probabilities of the other four relations can be expressed as follows: Is_diagnosed(x) has a given probability p_1 if Is_infectious(x) is true, and a (lower) probability p_2 otherwise.

Is_open(w) is deterministic, and it is true if neither there has been a diagnosed case at the workplace nor more than 1% of overall inhabitants have been diagnosed. In conditional probability logic, this is expressed as having probability 0 if

$$\exists_x(\text{Is\_diagnosed(x)} \wedge \text{Attends}(x, w)) \vee \mathbb{P}_x\,(\text{Is\_diagnosed}(x)) \geq 0.01$$

is true, and probability 1 otherwise.

Contact(x, y) models contact close enough to transmit the disease. It could be set at a probability p_3 if

$$\exists_w(\text{Is\_open}(w) \wedge \text{Attends}(x, w) \wedge \text{Attends}(y, w))$$

and at a (lower) probability p_4 otherwise. The values of p_3 and p_4 will be varied depending on the transmissivity of the disease and the social structure of the population.

(continued)

Finally, Is_Infected(x) can now be seen as a deterministic dependency, with probability 1 if

$$\exists_y(\text{Contact}(x, y) \wedge \text{Is\_Infectious}(y))$$

and probability 0 otherwise.

4.2.3 Type II Logics as a Query Language

Traditional Type II probability logics could be used as a powerful logical language to interact with statistical relational models. For instance, current query languages for statistical relational models do not make any use of their probabilistic semantics. While the details vary, queries are usually as ordinary quantifier-free formulas, possibly with additional formulas as "evidence", and by a command word to indicate the type of query desired, e. g. the conditional probability or the single most likely truth values given the evidence.

This could be supplemented very effectively with queries expressed in a full Type II probability logic. For instance one could directly query the independence of two propositions as an independence statement, one could ask threshold queries (Is this more than 50% likely?) or ask whether one event is more likely than the other.

In general, these queries could be much simpler than the exact marginal inference problem, which asks for the precise probability of a given proposition. A threshold query for instance could harness approximate inference with sharp bounds, and recent work by Rückschloß and Weitkämper (forthcoming) implies that conditional independence queries can be answered in linear time for some probabilistic logic programs.

4.3 Conclusion

Combining sophisticated logical reasoning with uncertainty has been an overarching objective of artificial intelligence research for the last three decades, and the long tradition of first-order logic, statistics and probability theory makes probabilistic reasoning a natural candidate for this integration. This gave rise to various logics of probability as well as to the field of statistical relational artificial intelligence, which combines relational logic with ideas from probabilistic graphical models. However, while one would assume probability logic to be a natural fit for the logical underpinnings of statistical relational formalisms, in fact they seem to have had little impact on each other.

In this contribution, I try to offer an explanation for the apparent divide between probability logic and statistical relational approaches, and hopefully open the prospect for harnessing probability logics for statistical relational artificial intelligence. Lessons can be learnt from the relationship between first-order logic and logic programming, and the taxonomy of probability logics developed by Halpern (1990) can guide the possible uses of those logics in statistical relational artificial intelligence. Independence statements emerge as a crucial expressive feature for making Type II probability logics usable in statistical relational contexts. When supporting independence statements, Type II probability logics can be used to provide a completion semantics for statistical relational models, and they can also open up new possibilities for a natural and highly expressive query language for statistical relational models. Type I probability logics on the other hand could take the role currently occupied by ordinary first-order logics in statistical relational models, as an expressive language in which to formulate predicate definitions or constraints on possible worlds.

By outlining possible applications to expressiveness, semantics and the design of query languages for statistical relational artificial intelligence, this chapter hopefully gives new impetus for the targeted study of first-order logics of probability and stimulate a new set of powerful tools in statistical relational methods and formalisms.

References

Abadi, M., and J.Y. Halpern. 1994. Decidability and Expressiveness for First-Order Logics of Probability. *Information and Computation* 112 (1): 1–36. https://doi.org/10.1006/inco.1994.1049.

Abiteboul, S., and V. Vianu. 1989. Fixpoint Extensions of First-Order Logic and Datalog-Like Languages. In *Proceedings of the Fourth Annual Symposium on Logic in Computer Science (LICS '89), Pacific Grove, California, USA, June 5-8, 1989*, 71–79. IEEE Computer Society. https://doi.org/10.1109/LICS.1989.39160.

Abiteboul, S., and V. Vianu. 1991. Datalog Extensions for Database Queries and Updates. *Journal of Computer and System Sciences* 43 (1): 62–124. https://doi.org/10.1016/0022-0000(91)90032-Z.

Azzolini, D., E. Bellodi, and F. Riguzzi. 2022. Statistical Statements inÂ Probabilistic Logic Programming. In *Logic Programming and Nonmonotonic Reasoning*, ed. G. Gottlob, D. Inclezan, and M. Maratea, 43–55. Cham: Springer International Publishing.

Bacchus, F. 1990. *Representing and Reasoning with Probabilistic Knowledge - A Logical Approach to Probabilities*. Cambridge: MIT Press.

Baral, C., M. Gelfond, and J.N. Rushton. 2009. Probabilistic Reasoning with Answer Sets. *Theory Pract Log Program* 9 (1): 57–144. https://doi.org/10.1017/S1471068408003645.

Carnap, R. 1950. *Logical Foundations of Probability*. Chicago: University of Chicago Press.

Clark, K.L. 1978 Negation as Failure. In *Logic and Data Bases*, ed. H. Gallaire, and J. Minker, 293–322. Boston: Springer.

Cozman, F.G., and D.D. Mauá. 2020. The Joy of Probabilistic Answer Set Programming: Semantics, Complexity, Expressivity, Inference. *International Journal of Approximate Reasoning* 125, 218–239. https://doi.org/10.1016/j.ijar.2020.07.004.

Ebbinghaus, H., and J. Flum. 1999. Finite Model Theory. *Perspectives in Mathematical Logic*. 2nd ed. Berlin: Springer.
Fierens, D., G.V. den Broeck, J. Renkens, D.S. Shterionov, B. Gutmann, I. Thon, G. Janssens, and L.D. Raedt. 2015. Inference and Learning in Probabilistic Logic Programs Using Weighted Boolean Formulas. *Theory and Practice of Logic Programming* 15 (3): 358–401. https://doi.org/10.1017/S1471068414000076
Halpern, J. Y. 1990. An Analysis of First-Order Logics of Probability. *Artificial Intelligence* 46 (3): 311–350. https://doi.org/10.1016/0004-3702(90)90019-V.
Ivanovska, M., and M. Giese. 2010. Probabilistic Logic with Conditional Independence Formulae. In *STAIRS 2010 - Proceedings of the Fifth Starting AI Researchers' Symposium, Lisbon, Portugal, 16-20 August, 2010*. Frontiers in Artificial Intelligence and Applications, ed. T. Ågotnes, vol. 222, 127–139. IOS Press. https://doi.org/10.3233/978-1-60750-676-8-127.
Jaeger, M. 1997. Relational Bayesian Networks. In *UAI '97: Proceedings of the Thirteenth Conference on Uncertainty in Artificial Intelligence, Brown University, Providence, Rhode Island, USA, August 1-3, 1997*, ed. D. Geiger, P.P. Shenoy, 266–273. Morgan Kaufmann. http://people.cs.aau.dk/~jaeger/publications/UAI97.pdf.
Keisler, H.J. 1985. Probability Quantifiers. In *Model-Theoretic Logics*. Perspectives in Mathematical Logic, 509–556. New York: Springer.
Koponen, V. 2020. Conditional Probability Logic, Lifted Bayesian Networks, and Almost Sure Quantifier Elimination. *Theoretical Computer Science* 848: 1–27. https://doi.org/10.1016/j.tcs.2020.08.006.
Koponen, V., and F. Weitkämper. 2023a. Asymptotic Elimination of Partially Continuous Aggregation Functions in Directed Graphical Models. *Information and Computation* 293: 105061. https://doi.org/10.1016/j.ic.2023.105061.
Koponen, V., and F. Weitkämper. 2023b. On the Relative Asymptotic Expressivity of Inference Frameworks. CoRR abs/2204.09457,. https://doi.org/10.48550/arXiv.2204.09457.
Lin, F., and Y. Zhao. 2004. Assat: Computing Answer Sets of a Logic Program by Sat Solvers. *Artificial Intelligence* 157 (1): 115–137. https://doi.org/10.1016/j.artint.2004.04.004. https://cse.hkust.edu.hk/faculty/flin/papers/assat-aij-revised.pdf.
Ognjanovic, Z., M. Raskovic, and Z. Markovic. 2016. *Probability Logics - Probability-Based Formalization of Uncertain Reasoning*. Berlin: Springer. https://doi.org/10.1007/978-3-319-47012-2.
Perovic, A., Z. Ognjanovic, M. Raskovic, and Z. Markovic. 2008. A Probabilistic Logic with Polynomial Weight Formulas. In *Foundations of Information and Knowledge Systems, 5th International Symposium, FoIKS 2008, Pisa, Italy, February 11-15, 2008, Proceedings*. Lecture Notes in Computer Science, ed. S. Hartmann, and G. Kern-Isberner, vol. 4932, 239–252. Berlin: Springer. https://doi.org/10.1007/978-3-540-77684-0_17.
Poole, D. 1993. Probabilistic Horn Abduction and Bayesian Networks. *Artificial Intelligence* 64 (1): 81–129. https://doi.org/10.1016/0004-3702(93)90061-F.
Raedt, L.D., K. Kersting, S. Natarajan, and D. Poole. 2016. Statistical Relational Artificial Intelligence: Logic, Probability, and Computation. *Synthesis Lectures on Artificial Intelligence and Machine Learning*. San Rafael: Morgan & Claypool Publishers. https://doi.org/10.2200/S00692ED1V01Y201601AIM032.
Riguzzi, F. 2023. *Foundations of Probabilistic Logic Programming: Languages, Semantics, Inference and Learning*. 2nd ed. Gistrup: River Publishers.
Riguzzi, F., and T. Swift. 2018. A Survey of Probabilistic Logic Programming. In *Declarative Logic Programming: Theory, Systems, and Applications*, ed. M. Kifer, and Y.A. Liu, 185–228. New York: ACM/Morgan & Claypool. https://doi.org/10.1145/3191315.3191319.
Sato, T. 1995. A Statistical Learning Method for Logic Programs with Distribution Semantics. In *Logic Programming, Proceedings of the Twelfth International Conference on Logic Programming, Tokyo, Japan, June 13-16, 1995*, ed. L. Sterling, 715–729. Cambridge: MIT Press.
Truszczynski, M. 2018. *An Introduction to the Stable and Well-Founded Semantics of Logic Programs*, 121–177. New York: Association for Computing Machinery and Morgan & Claypool. https://doi.org/10.1145/3191315.3191318.

Warren, D.S., and M. Denecker. 2023. A Better Logical Semantics for Prolog. In *Prolog: The Next 50 Years*. Lecture Notes in Computer Science, ed. D.S. Warren, V. Dahl, T. Eiter, M.V. Hermenegildo, R. Kowalski, and F. Rossi, vol. 13900, 82–92. Berlin: Springer. https://doi.org/10.1007/978-3-031-35254-6_7.

Weitkämper, F. 2022. Statistical Relational Artificial Intelligence with Relative Frequencies: A Contribution to Modelling and Transfer Learning Across Domain Sizes. 2202.10367.

Chapter 5
An Overview of the Generalization Problem

Francesco Facciuto

Abstract The *generalization problem* is examined, providing motivations for the introduction of Information Theory tools. The proposed analysis will suggest the following general picture: although Machine Learning techniques are classically conceptualized as *curve-fitting* methods, protocols that generalize effectively respond to a partially different logic by implementing *coarse-graining* procedures. In summary, the first paradigm provides quantitative guarantees about generalization when the hypothesis space is predetermined. However, this approach inherits some relevant limitations. Consequently, the *compression-generalization* trade-off can be adopted as an alternative. Without the compression process, the effectiveness of artificial intelligence protocols—particularly in the realm of Deep Learning—would likely be far less impressive than what we currently observe.

Keywords Learning · Artificial intelligence · PAC bounds · Information · Large-scale problems · Compression and generalization

5.1 Framing the Problem

We examine the problem of determining the conditions under which a recognized pattern can be considered representative of the studied phenomenon, extending beyond the specific dataset used for inference. This leads us to the *generalization problem*, which arises in both theoretical and practical contexts. Generalization appears to be a fundamental challenge every cognitive system faces: an intelligent agent does not merely store all the available information but discards irrelevant details, constructing a simplified representation that can adapt to various situations. A similar challenge exists within the realm of Machine Learning, where the prominent goal is to develop protocols that provide accurate predictions for data

F. Facciuto (✉)
Università degli Studi di Milano, Milan, Italy
e-mail: francesco.facciuto@unimi.it

H. Hosni, J. Landes (eds.), *Perspectives on Logics for Data-driven Reasoning*, Logic, Argumentation & Reasoning 35,
https://doi.org/10.1007/978-3-031-77892-6_5

points outside the training set. This specific context will be explored to obtain a rigorous characterization of generalization, which can be adaptable to different domains.

Machine Learning and Pattern Recognition Digital computers are designed to be programmed by humans, so they must follow instructions precisely. However, we now have a different approach for enabling general-purpose systems to perform specific computational tasks: learning from data. Machine Learning refers to a broad set of tools designed to solve data inference problems. In this discussion, we adopt a quite narrow perspective, excluding both generative modeling and symbolic approaches, and instead consider algorithms that learn a function from a predefined class. Algorithms that perform this task using labeled data are said to operate in a *supervised* mode (Mehta et al. 2019; Murphy 2013). For simplicity, we will always limit our focus to this case. In a nutshell, the starting point consists of an input-output database

$$\mathcal{D} = \{(x_1, y_1), (x_2, y_2), ..., (x_n, y_n)\},$$

where we want to reconstruct the rule $y = f(x)$, or at least an approximation. Once learned, these rules can be applied to tasks such as document or image classification (Baldi 2021; Goodfellow et al. 2016; Murphy 2013), stock price prediction (Soni et al. 2022), or diagnosing diseases from patient records (Sidey-Gibbons and Sidey-Gibbons 2019). To achieve this goal, a predefined architecture

$$\mathcal{A}_\theta : \mathcal{D} \longmapsto f_\theta$$

is selected, with free parameters $\theta \in \Theta$ to be determined. The general recipe consists of three steps. First, $\mathcal{D}$ is partitioned into two subsets distributed according to the same probability distribution, $\mathcal{D}_{train}$ and $\mathcal{D}_{test}$. Second, during the *training phase*, the algorithm processes $\mathcal{D}_{train}$, and an optimization procedure is used to determine

$$\theta^* := \arg\min_{\theta \in \Theta} C(\theta; \mathcal{D}_{train}), \tag{5.1}$$

where the predefined *cost function* $C(\theta; \tilde{\mathcal{D}})$ quantifies the accuracy of f_θ on $\tilde{\mathcal{D}} \subseteq \mathcal{D}$. Finally, the ability of f_{θ^*} to generalize beyond $\mathcal{D}_{train}$ is evaluated on $\mathcal{D}_{test}$, and the primary goal is for f_{θ^*} to achieve a low *test cost* $C(\theta^*; \mathcal{D}_{test})$. In this regard, the *overfitting* problem can come into play: if the trained model f_{θ^*} is too specialized on $\mathcal{D}_{train}$, it may have captured the noise in the data, thus reducing its predictive performance on $\mathcal{D}_{test}$. In other words, as $C(\theta; \mathcal{D}_{train})$ decreases with varying θ, $C(\theta; \mathcal{D}_{test})$ may increase. To mitigate this, additional steps can be taken between the training and testing phases by proceeding with *fine-tuning* over some *hyper-parameters*, often making decisions based on intuitions and previous experiences (Ng 2017). A standard technique to prevent overfitting consists of equipping the cost function with a *regularization term* (Bishop 2007; Murphy 2013), thus replacing

(5.1) with

$$\theta^* := \arg\min_{\theta \in \Theta} \left\{ C(\theta; \mathcal{D}_{train}) + \gamma ||\theta|| \right\}, \tag{5.2}$$

where $\gamma \geq 0$ and $|| \cdot ||$ denotes some norm. Choosing the hyperparameter γ is equivalent to introducing a penalty that constrains the magnitude of θ. While these techniques can be effective in practice, they lack a complete theoretical justification, and the question *How is it possible to generalize well?* remains only partially answered. The following section discusses a classic attempt to address this gap.

5.2 Statistical Learning Theory

Classical Statistical Learning Theory (Valiant 1984; Vapnik 2013; Vapnik and Vapnik 1998) allows us to formulate the generalization problem in a mathematically precise way, by providing the trade-off between the *accuracy* and *complexity* of the function f chosen from a functional space $\mathcal{F}$ (Mohri et al. 2012). Intuitively, we expect that as the size of the dataset $\mathcal{D}$ increases, the function selected by the learning algorithm $\mathcal{A}$ will become more accurate. This perspective is supported by the fact that traditional machine learning protocols are *universally consistent*, i.e. every function f can be learned exactly in the limit of infinite data (Shalev-Shwartz and Ben-David 2014). However, universal consistency provides no guarantee regarding the accuracy achievable starting from a finite dataset. In practice, a consistent algorithm may not outperform a non-consistent one, and the phenomenon of *slow rates* may come into play. The *No Free Lunch Theorem* suggests that, for any learning algorithm, there are problems for which the learning rate is arbitrarily slow (Murphy 2013; Shalev-Shwartz and Ben-David 2014). This highlights the need to consider the size of the available dataset as a key parameter in our analysis. To clarify this, we will focus on the supervised learning setting (Murphy 2013; Vapnik 2013), and for simplicity, restrict our discussion to *binary classification* tasks.

Minimizing the Expected Error Let us consider: a random variable X distributed according to a probability distribution P_X; a supervisor that returns an output y according to a distribution $P_{Y|X}$; an algorithm that returns a function $f \in \mathcal{F}$, selected to emulate the supervisor's response. In general, the distribution $P_{X,Y}$ is unknown. The algorithm takes as input a training set $\mathcal{D} = \{(x_1, y_1), \cdots, (x_n, y_n) : (x_i, y_i) \in \mathcal{X} \times \mathcal{Y}\}$ consisting of *independently and identically distributed* instances sampled from $P_{X,Y}$. The quality of f must be evaluated on instances distributed as $P_{X,Y}$, but not necessarily included in $\mathcal{D}$. In other words, we want to ensure that f is not specialized just on the observed data but captures the underlying regularity of the problem instead. Given an occurrence for X, it is to be determined a rule that associates the corresponding value in $\mathcal{Y}$. We seek to learn the optimal $f : \mathcal{X} \rightarrow \mathcal{Y}$, i.e. the one that minimizes the *Expected Error*

$$R[f] := \int \text{err}[f(x), y] dP_{X,Y}$$

over the class of functions $\mathcal{F}$. Since the distribution $P_{X,Y}$ is unknown, we can not directly minimize the functional $R[f]$.

Induction Principle To avoid the difficulty mentioned above, we consider instead of $P_{X,Y}$ the empirical counterpart $\hat{P}_{X,Y}$, which is the distribution concentrated on the available data in $\mathcal{D}$. Through this particular choice, the functional $R[f]$ is replaced by the *Empirical Risk*

$$\hat{R}_n[f] := \frac{1}{n} \sum_{i=1}^{n} \text{err}[f(x_i), y_i],$$

so that the algorithm $\mathcal{A}$ will proceed by minimizing $\hat{R}[f]$. However, this procedure does not provide the necessary guarantees. To see this, it is sufficient to consider the following situation: if we consider the entire space of binary functions $\mathcal{F} := \{f : \mathcal{X} \to \{0, 1\}\}$, our algorithm could in principle select one of the $2^{|\mathcal{X}|-n}$ possible rules for which $\hat{R}_n[f] = 0$, without being able to establish a priori which of these best generalizes on the new instances. In other words, by minimizing the Empirical Risk, we could select a function in the *overfitting regime*. As we will see, the classic strategy to overcome this difficulty consists of restricting the space of functions $\mathcal{F}$. To introduce this point, let us start by considering how this restriction acts at the level of the Expected Error. It is worth keeping in mind that the space $\mathcal{F}$ is not necessarily given a priori, so this evaluation is significant per sé.

Classical Error Decomposition We can evaluate how a gradual narrowing of the functional space—from $\mathcal{F}$ to one of its subsets $\mathcal{F}_\delta$—affects the performance of the learning algorithm. Let us consider a function $f_c \in \mathcal{F}_\delta$ by assuming it has been chosen after the training phase, evaluating the difference between $R[f_c]$ and the optimal expected error $\inf_{f \in \mathcal{F}} R[f]$. This difference quantifies the quality of the selected function. By adding and subtracting equal terms, we immediately obtain the following equality:

$$\begin{aligned} R[f_c] - \inf_{f \in \mathcal{F}} R[f] = & \quad R[f_c] - \hat{R}_n[f_c] \\ + & \quad \hat{R}_n[f_c] - \inf_{f \in \mathcal{F}_\delta} \hat{R}_n[f] \\ + & \quad \inf_{f \in \mathcal{F}_\delta} \hat{R}_n[f] - \inf_{f \in \mathcal{F}_\delta} R[f] \\ + & \quad \inf_{f \in \mathcal{F}_\delta} R[f] - \inf_{f \in \mathcal{F}} R[f] \end{aligned}$$

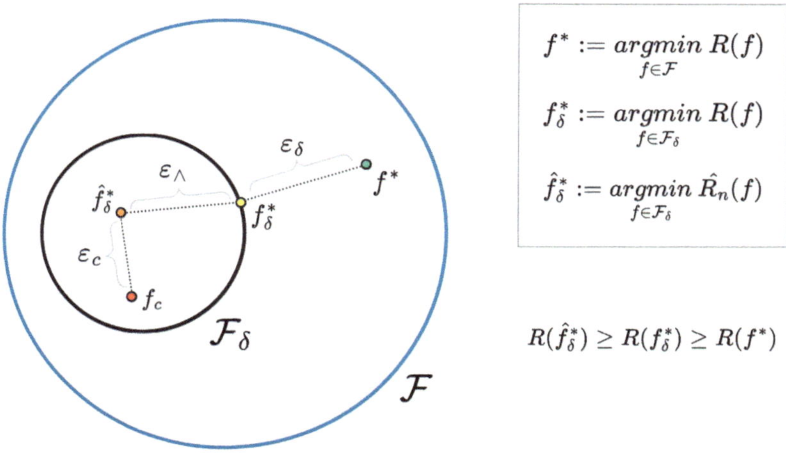

Fig. 5.1 The Classical Error Decomposition is schematically illustrated by considering a restriction of the assigned functional space, from $\mathcal{F}$ to $\mathcal{F}_\delta$

By applying a straightforward upper bound, it is easy to recognize the following relationship, as also schematically depicted in Fig. 5.1:

$$
\begin{aligned}
R[f_c] - \inf_{f\in\mathcal{F}} R[f] \leq\ & \hat{R}_n[f_c] - \inf_{f\in\mathcal{F}_\delta} \hat{R}_n[f] && (\varepsilon_c)\\
+\ & 2 \sup_{f\in\mathcal{F}_\delta} \left| R[f] - \hat{R}_n[f] \right| && (\varepsilon_\wedge)\\
+\ & \inf_{f\in\mathcal{F}_\delta} R[f] - \inf_{f\in\mathcal{F}} R[f] && (\varepsilon_\delta)
\end{aligned}
$$

Let us briefly examine the three contributions considered above. The first one—denoted as ε_c—reflects the possibility that the algorithm may not solve the empirical risk optimization exactly, hence returning a suboptimal function, in some cases corresponding to a local minimum. While significant, this optimization challenge is unrelated to the problem of generalization: the gap remains non-zero only if f_c fails on the training set already. Therefore, we will omit any further discussion on it. The third one—ε_δ—explains the error arising from the narrowing of the functional space, which manifests as a systematic bias and is independent of the dataset. The second one—$\varepsilon_\wedge$—corresponds to a uniform control on $\mathcal{F}_\delta$ of the difference between the Expected Error R and the Empirical Risk $\hat{R}_n$. This contribution is the only one related to the problem of generalizing from a finite dataset and it is the only one that can be controlled by the dataset size. This point can be analyzed by adopting the classic *Probably Approximately Correct* (PAC) learning framework, as briefly discussed in the following paragraph.

Capacity Measures and PAC Bounds By introducing a measure of complexity for the functional space $\mathcal{F}$, it is possible to obtain a uniform control of the difference between Empirical Risk and Expected Error, establishing a convergence rate that depends on the size of the dataset. The main idea is to exploit the concentration in probability for $\hat{R}_n[f]$ around $R[f]$ due to the *Chernoff-bound* effect [Appendix 1]. If $\mathcal{F}$ has finite cardinality, it is possible to demonstrate that the following inequality holds—for any $f \in \mathcal{F}$—with probability $1 - \delta$:

$$|R[f] - \hat{R}_n[f]| \leq \sqrt{\frac{\log(2|\mathcal{F}|/\delta)}{2n}}. \tag{5.3}$$

This inequality establishes a relation between the *generalization gap* $|R[f] - \hat{R}_n[f]|$ and the function complexity measured by the *capacity* $|\mathcal{F}|$, controlled by the dataset size n. In other words, we can estimate the number of instances required to achieve a high probability control over the generalization gap, and a non-trivial trade-off comes into play. On the one hand, we would like to work with a large class of functions to allow for a small Empirical Risk, i.e. minimizing ε_δ. On the other hand, the functional class should be small enough to control the generalization gap $\varepsilon_\wedge$. Optimizing this trade-off is a well-defined problem also for other capacity measures, e.g. when $\mathcal{F}$ is an infinite set [Appendix 2]. At any rate, the following three remarks are essential and of entirely general scope.

- First, $\mathcal{F}$ has a central role. It represents the *background knowledge* a priori available and corresponds to some hypothesis of regularity about the data. However, it is not necessarily the case that our knowledge of the problem is sufficient to select $\mathcal{F}$ such that it is both relevant—i.e. containing the correct function—and has a limited capacity. Methodologically, $\mathcal{F}$ constitutes a *spurious* object in a properly inductive scenario.
- Second, the inequality (5.3) is independent of the probability distribution P. In other words, the particular problem at hand is not considered, giving us a *worst-case* result that does not exploit possible distribution regularities. This perspective does not seem satisfactory: we would like a bound that depends on the complexity of P, establishing some classification that distinguishes tractable distributions from intractable ones.
- Finally, the derived bound does not depend on the learning algorithm $\mathcal{A}$ and appears unable to account for the results achieved in recent developments in Machine Learning. Deep Learning models—despite their vast number of parameters and the resulting high $\mathcal{F}$-capacity—demonstrate the ability to learn complex problems with far less data than the bound would suggest. In many cases, the number of parameters exceeds the dataset size by orders of magnitude, making the classical control over the generalization gap meaningless.

The third point is particularly relevant when considering that $\mathcal{F}$ typically has infinite cardinality in applications. In this context, the *curse of dimensionality* immediately becomes apparent. To clarify this point, let us consider $\mathcal{F}$ as a set of functions f :

$\mathcal{X} \sim \mathbb{R}^d \to \{0, 1\}$ and estimate its size by restricting it to a compact domain, $[0, 1]^d$. By dividing the space into small d-dimensional cubes of side length ε, we obtain a quantized version $\mathcal{X}_\varepsilon$ of $\mathcal{X}$, containing ε^{-d} distinct elements. The cardinality of $\mathcal{F}_\varepsilon$ is then $2^{\varepsilon^{-d}}$, and the superexponential growth with respect to dimension d indicates that high-dimensional problems can be unlearnable in the regime of finite data (see Appendix 2 for further details). In summary, the three observations mentioned above highlight the necessity for a shift in perspective.

5.3 Information Bounds

Let us consider the binary classification problem for the space $\mathcal{X} = \mathcal{X}_1 \times \mathcal{X}_2 \times ... \times \mathcal{X}_m$ of binary strings of length m. The set of the possible classification rules $f : \mathcal{X} \to \mathcal{Y} \sim \{0, 1\}$ has cardinality $2^{|\mathcal{X}|} = 2^{2^m}$. Without further assumptions, identifying the correct f is an intractable problem: given the relation (5.3), the generalization gap results controlled by ε only if n is at least of order $O(2^m/\varepsilon^2)$. Instead of considering a predefined subset $\mathcal{F} \subset \mathcal{Y}^{\mathcal{X}}$, we focus on functions defined over an *effective* subset $\mathcal{X}_{\text{eff}} \subset \mathcal{X}$, with probability approaching 1. To begin, let us explore this possibility by removing low-probability patterns.

Removing Negligible Patterns We discuss a rigorous generalization-gap bound that depends on the complexity of the input variable X, as quantified by the Shannon Entropy $H[X]$. This preliminary result is based on characterizing the set of *non-negligible* patterns, offering initial insight into how a more compressed representation can improve the effectiveness of the adopted inductive protocol. The core idea is to define an input space $\mathcal{X}_\sigma$ limited to instances that occur with a probability exceeding a fixed threshold σ. By this cardinality reduction, the space of possible functions shrinks, and the generalization gap can be consequently controlled via a better bound. Intuitively, the more instances occur with different probabilities, the more it becomes possible to exclude a significant portion of the inputs. This reduction can be evaluated by considering the ratio

$$\frac{H[X]}{\log \frac{1}{\sigma}},$$

where the denominator corresponds to the entropy of a uniform random variable counting $1/\sigma$ elements. This ratio comes into play in relation to the error ε. In Appendix 3 we show that the generalization gap is controlled by ε with probability $1 - \delta$ and uniformly in $f \in 2^{\mathcal{X}}$ if n scales as

$$n > \frac{2^{\frac{6H[X]}{\varepsilon}} + \log \frac{2}{\delta}}{\varepsilon^2}. \tag{5.4}$$

Although the relation (5.4) is independent of the function class $\mathcal{F}$, it remains to be discussed to what extent the obtained bound provides us with an adequate estimate of the dataset size required for generalization. In this regard, two remarks are needed.

- The bound (5.4) provides a first insight into how a *more compressed* representation $\mathcal{X}_{\text{eff}}$ of $\mathcal{X}$ can enhance the efficacy of the inductive protocol adopted: reducing $H[X]$ decreases exponentially the number of data required to achieve the desired control on the generalization gap. This point will be clarified further.
- On the other hand, the exponential dependence on $1/\varepsilon$ in (5.4) introduces a significant difficulty: for small ε the dataset size n required for learning becomes unreasonably large, regardless of how small the entropy $H[X]$ is. Remarkably, factor $1/\varepsilon$ appears by construction from the definition of the threshold σ, used to define the set $\mathcal{X}_\sigma$ of *non-negligible* patterns. In other words, it seems to be unavoidable without additional assumptions [Appendix 3].

In the following paragraphs, a different perspective is considered, for which the notion of information plays a more direct role. If the Large Scale Learning regime—where the number of training examples and the input dimension are both huge—is considered, it is possible to introduce a typical-case bound obtained by restricting the probability distributions to a suitable class. In this regard, N. Tishby has proposed this approach to justify Deep Learning's ability to generalize (Tishby and Zaslavsky 2015; Shwartz-Ziv and Tishby 2017; Shwartz-Ziv 2022).

Some Elementary Remarks on Shannon Entropy If a stochastic process satisfies certain assumptions, the probability $p(x_1, x_2, ..., x_m)$ of observing a sequence $(x_1, x_2, ..., x_m)$ is arbitrarily close to $2^{-mH[X]}$ for most of the sequences observed as m grows. Moreover, the space of all sequences can be divided into *typical* and atypical, where the probability of observing atypical ones approaches 0 when m is large. Let us start discussing this point in the simplest case, i.e. by assuming that the probability P_X is factorized as

$$P_X = \prod_{i=1}^{m} P_{X_i},$$

where P_{X_i} refers to a single bit of X and $P_{X_j} = P_{X_k}$ for all j and k. The normalized sum of independent terms concentrates—for large m—around the mean:

$$-\frac{1}{m}\log p(x_1, x_2, ..., x_m) = -\frac{1}{m}\sum_{i=1}^{m}\log p(x_i) \approx H[X_k], \tag{5.5}$$

$$H[X_k] := -\sum_{x_k \in \mathcal{X}_k} p(x_k)\log p(x_k).$$

In a nutshell, the *Shannon Entropy* $H[X]$ dominates the effective size of $\mathcal{X}$, i.e.

$$|\mathcal{X}_{\text{eff}}| \approx 2^{mH[X_k]} = 2^{H[X]}.$$

This circumstance arises for a wider class of processes, e.g. when the Shannon-McMillan-Breiman theorem holds. At any rate, these simple remarks are sufficient to show how the *compression-generalization* trade-off can be adopted as a general paradigm: if only typical patterns are considered—each one with probability $2^{-H[X]}$—a coarser representation comes into play without the need for introducing capacity measures on a predetermined $\mathcal{F}$. Although the extent to which typicality is a desirable property remains to be determined, the following statement will be examined in the next step: more compressed representations allow us to control the generalization gap better.

Compression in Representation Machine Learning protocols rely on internal representations to enhance their performance (Baldi 2021; Bengio et al. 2013; Goodfellow et al. 2016). While determining an appropriate representation can be essential for numerous practical domains, the precise criteria for establishing a satisfactory one continue to pose significant challenges (Schölkopf et al. 2021). From a general viewpoint, *representation learning* refers to transforming the high-dimensional input space $\mathcal{X}$ into a lower-dimensional one $\mathcal{T}$. In this regard, most of the input's entropy $H[X]$ turns out to be irrelevant, and the main challenge consists of extracting the relevant features to construct T. Schematically, the learning apparatus can be seen as a Markov chain

$$Y \longleftrightarrow X \longrightarrow T \longrightarrow \hat{Y}, \tag{5.6}$$

where the representation T is selected via the training phase to obtain the assigned label $\hat{Y}$ as close as possible to the correct label Y. This formalism allows us to frame the problem of effective representation in probabilistic terms while necessitating the introduction of suitable counterparts for $R[f]$ and $\hat{R}[f]$, which—instead of a deterministic f—consider probability distributions. Mutual information will play this role [Appendix 4]. Moreover, two different and separate components can be considered, although these may occur interdependently during the training phase:

- The *feature extraction* component corresponds to the encoding procedure $X \longrightarrow T$, for which the *encoder* $p(t|x)$ must be determined. The mutual information $I(X : T)$ quantifies the level of information shared between representations X and T, reducing to $H[T]$ when T is characterized deterministically by X.
- The *label estimation* component corresponds to the decoding procedure $T \longrightarrow \hat{Y}$, for which the *decoder* $p(y|t)$ should be as close as possible to the *optimal decoder*

$$p_{\text{opt}}(y|t) := \sum_{x \in \mathcal{X}} p(y|x)p(x|t). \tag{5.7}$$

The discussion so far has been concerned only with this second stage.

Let us assume that the representation T is fixed and the encoder $p(t|x)$ is assigned. The mutual information $I(Y : T)$ can be interpreted as an accuracy index, as justified by the following relation [Appendix 4]:

$$\begin{aligned}\mathbb{E}_{p(x,t)}\big[||p(y|x) - p(y|t)||_1^2\big] &\leq \frac{1}{2\ln 2}\mathbb{E}_{p(x,t)}\big[D[p(y|x)||p(y|t)]\big] \\ &= \frac{1}{2\ln 2}\big\{I(X:Y) - I(Y:T)\big\}. \end{aligned} \tag{5.8}$$

The mutual information $I(X : Y)$ does not depend on the particular learning procedure. Consequently, to obtain an optimal decoder $p(y|t)$ nearer to the original distribution $p(y|x)$, $I(Y : T)$ should increase regardless of the learning protocol adopted. To clarify how different representations can intervene in improving (5.4), let us consider $\mathcal{T}$ as a partition of $\mathcal{X}$ for which a *homogeneity condition* holds: all instances assigned to the same $t \in \mathcal{T}$ share the same label y. Assuming (5.7), the Expected Error and Empirical Risk can be re-written as

$$\begin{aligned} R[f] &= \int \mathrm{err}[f(x), y]p(x, y)dxdy = \int \mathrm{err}[f(x), y] \int p(x, y, t)dxdydt \\ &= \int \mathrm{err}[f(t), y] \left\{ \int p(y|x)p(x|t)dx \right\} p(t)dydt \\ &= \int \mathrm{err}[f(t), y]p(t, y)dtdy, \\ \hat{R}_n[f] &= \frac{1}{n}\sum_{\mathcal{D}} \mathrm{err}[f(x_i), y_i] = \frac{1}{n}\sum_{\mathcal{D}} \mathrm{err}[f(t_i), y_i]. \end{aligned}$$

At this point, the argument employed above can be directly applied by replacing $H[X] \longmapsto H[T]$ in (5.4). However, observing that Y is independent of T given X and weakening the homogeneity by introducing a soft partition via a non-deterministic $p(t|x)$, we expect that (5.4) can be re-written as

$$n \sim \frac{2^{\frac{6I(X:T)}{\varepsilon}} + \log\frac{2}{\delta}}{\varepsilon^2}. \tag{5.9}$$

The relation (5.9) suggests that if the accuracy $I(Y : T)$ remains fixed, we expect that a more compressed representation $\mathcal{T}$—as quantified by the mutual information $I(X : T)$—improves the learning performance. At this point, two comments are needed.

- Although estimate (5.9) has not been rigorously proved, we will find a similar result when considering the typicality argument. Again, we will assume T fixed and neglect *non-typical* patterns instead of negligible ones.
- The informational quantities considered so far are computable only if $p(x, y, t)$ is available. On the other hand, assuming only the empirical distribution $\hat{p}(x, y)$ and the encoder $p(t|x)$ are available, it is possible to show that

$$\mathbb{E}_{\hat{p}(x,t)}\big[||\hat{p}(y|x) - p(y|t)||_1^2\big] \leq \frac{1}{2\ln 2}\mathbb{E}_{\hat{p}(x,t)}\big[D[\hat{p}(y|x)||p(y|t)]\big]$$
$$= \frac{1}{2\ln 2}\big\{I(X:Y) - \hat{I}(Y:T)\big\},$$

where the empirical quantity $\hat{I}(Y : T)$ plays the role of the Empirical Risk in the classical PAC bounds setting via (Shamir et al. 2010; Tishby and Zaslavsky 2015)

$$I(Y:T) \leq \hat{I}(Y:T) + O\left(\sqrt{\frac{2^{I(X:T)}}{n}}\right). \tag{5.10}$$

If (5.3) is considered, the analogy with the relation above is clear, and selecting the decoder $p(y|t)$ to maximize the empirical quantity $\hat{I}(Y : T)$ consequently becomes a good strategy to maximize $I(Y : T)$. Whereas control in (5.3) improves only logarithmically in $|\mathcal{F}|$, a reduction in $I(X : T)$ provides an exponential improvement in (5.10).

The compression index $I(X : T)$ can always be reduced by ignoring some details in $\mathcal{X}$. An essential point is to establish a link between compression level and its effects in terms of accuracy. Therefore, if the encoder $p(t|x)$ is not provided a priori but learned during the training, additional constraints on T are needed: the decoder $p(y|t)$ must be near to the original one $p(y|x)$. In other terms, the mutual information $I(Y : T)$—as it appears in (5.8)—should be maximized, as discussed below.

Information Bottleneck Principle Let T be a random variable to be considered as in the learning apparatus (5.6). The encoder $p(t|x)$ induces a soft partition of $\mathcal{X}$, for which each value $x \in \mathcal{X}$ is associated with all $t \in \mathcal{T}$ via $p(t|x)$. In our intentions, T should be a compressed representation of X so that $|\mathcal{T}| \leq |\mathcal{X}|$ and $p(t|x)$ results appropriately concentrated. In this regard, $I(T : X)$ provides a compactness index for the representation T, although additional prescriptions are needed to evaluate the representation adequacy. In this regard, the *rate-distortion theory* offers a first type of *relevance criteria* by providing a metric

$$\Delta : \mathcal{X} \times \mathcal{T} \longrightarrow \mathbb{R}^+,$$

assuming that smaller values of $\Delta(x,t)$ imply a better representation $\mathcal{T}$ (Cover and Thomas 2012; Cecconi et al. 2013). In a nutshell, the partition of $\mathcal{X}$ induced by the encoder $p(t|x)$ corresponds to an expected distortion $\mathbb{E}_{p(x,t)}[\Delta(x,y)]$ and a trade-off with the representation compactness is established via the constrained optimization of $I(X:T)$, so that

$$R(D) := \min_{\{p(t|x):\mathbb{E}_{p(x,t)}[\Delta(x,y)]\leq D\}} I(X:T).$$

This problem can be re-formulated by introducing the Lagrange multiplier β and minimizing the Lagrangian functional

$$\mathcal{L}[p(t|x)] := I(X:T) + \beta\mathbb{E}_{p(x,t)}[\Delta(x,y)] \tag{5.11}$$

under the additional constraint $\sum_t p(t|x) = 1$ for all $x \in \mathcal{X}$. The variational problem $\frac{\delta\mathcal{L}}{\delta p(t|x)} = 0$ admits the implicit solution

$$\begin{cases} p_\beta(t|x) = \dfrac{p_\beta(t)}{Z(x,\beta)} e^{-\beta\Delta(x,t)} \\ p_\beta(t) = \displaystyle\sum_{x\in\mathcal{X}} p_\beta(t|x)p(x) \end{cases}$$

where $Z(x,\beta)$ is a normalization function and $\beta \geq 0$ satisfies

$$\beta = -\frac{\delta R}{\delta D}.$$

$R(D)$ is defined with respect to a fixed set of representatives $\mathcal{T}$: given different values of $|\mathcal{T}|$, different distortion matrices $[\Delta(x,t)]_{x,t}$ are defined, changing $R(D)$. The rate-distortion $R(D)$ provides a monotonic convex curve with slope $-\beta$ in the (D, I)-plane (Cover and Thomas 2012), to be determined numerically by choosing different values of β and applying the iterative *Blahut-Arimoto* procedure (Slonim 2002). Theoretically, when β increases, minimizing the expected distortion becomes more and more relevant in (5.11), up to $\mathbb{E}_{p(x,t)}[\Delta(x,y)] = 0$ when $I(T:X) = H[X]$. On the other hand, the compression level matters more and more when β decreases, up to $I(T:X) = 0$ for a certain value of the distortion onward. Two comments are needed.

- First, the curve obtained provides two regions in the (D, I)-plane. The region above the curve corresponds to all the achievable distortion-compression pairs, while the region below is in principle not achievable. Given a distortion-compression pair (D^*, I^*), there exists an encoder $p(t|x)$ for which $I(X:T) = I^*$ and $\mathbb{E}_{p(x,t)}[\Delta(x,y)] = D^*$ if and only if (D^*, I^*) is above the curve, on which the optimal encoder is placed.

- Second, the characterization of $R(D)$ relies on the spurious element Δ, which introduces an arbitrary choice. In this regard—for a given $p(x)$—different choices of Δ will yield different results and alternative rate-distortion curves, so selecting the appropriate one is far from trivial in many practical applications. In other words, the main drawback of the rate-distortion approach consists of considering Δ as a part of the problem.

The Information Bottleneck Principle (Tishby et al. 2001; Slonim and Tishby 1999; Gilad-Bachrach et al. 2003) provides an alternative to overcome the above mentioned difficulties: instead of considering only $p(x)$, also $p(x, y)$ comes into play to determine a compressed representation of X that preserves the information on Y as high as possible. Starting from the learning apparatus (5.6), the optimal encoder is characterized as

$$\min_{p(t|x)} I(X:T) - \beta I(T:Y), \tag{5.12}$$

where β corresponds to a resolution parameter for T: small β implies more compression at the expense of informativeness, while bigger β corresponds to finer granularity in T. The variational problem (5.12) admits the implicit solution

$$\begin{cases} p_\beta(t|x) = \dfrac{p(t)}{Z(x,\beta)} e^{-\beta D[p(y|x)||p_\beta(y||t)]} \\ p_\beta(t) = \displaystyle\sum_{x \in X} p_\beta(t|x) p(x) \\ p_\beta(y|t) = \displaystyle\sum_{x \in X} p(y|x) p_\beta(x|t) \end{cases}$$

where the KL-divergence $D[p(y|x)||p(y||t)]$ assumes the role of the distortion measure (Peter Harremoes and Naftali Tishby 2007) and

$$\beta = \frac{\delta I(X:T)}{\delta I(T:Y)}.$$

Similarly to the case of distortion, the three equations above determine self-consistently the optimal solution (Slonim 2002). In contrast to rate distortion theory—where the representative selection is a separate problem—the optimization is also over the cluster representatives $p(y|t)$. If the decoder $p_\beta(y|t)$ is near to $p(y|x)$, the encoder $p_\beta(t|x)$ is concentrated around the good t. In this regard, we have a correspondence with a *clustering problem* (Murphy 2013; Slonim 2002; Slonim and Tishby 1999).

Mimicking what has already been discussed for $R(D)$, the *Information Bottleneck curve* is obtained by solving (5.12) as $\beta \geq 0$ varies and plotting the mutual information pair $(I_\beta(X:T), I_\beta(Y:T))$ given the optimal encoder $p_\beta(t|x)$. This

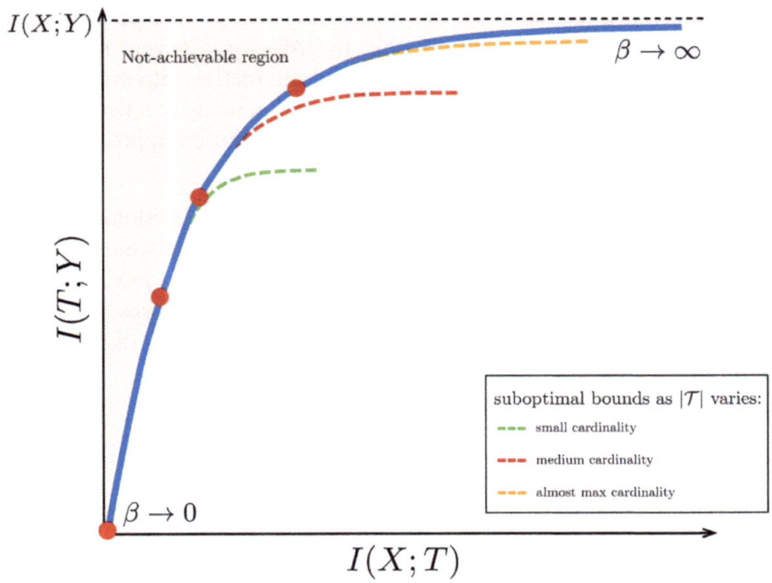

Fig. 5.2 *Information Bottleneck curve in the Information Plane.* As β varies, a monotonic concave curve is provided. Only the region below the curve is achievable. Fixing the representation cardinality $|\mathcal{T}|$ provides sub-optimal curves. The optimal representation T for $p(x, y)$ is characterized by the compression-accuracy couple, regardless of the learning procedure adopted

curve in the *Information Plane*—i.e. the plane with $I(X : T)$ and $I(Y : T)$ as the x-axis and y-axis respectively (Tishby et al. 2001; Tishby and Zaslavsky 2015; Shwartz-Ziv and Tishby 2017)—is concave and monotonically increasing, as shown in Fig. 5.2. Theoretically, given the distribution $p(x, y)$, the optimal representation T is completely determined via a general principle without introducing additional spurious elements such as the distortion measure. On the other side, the explicit calculation of the solution is generally prohibitive, with some exceptions (Chechik et al. 2003; Goldfeld and Polyanskiy 2020). Note that (5.12) does not pertain to a particular learning procedure but refers to the possibility of an appropriate representation in general. This high-level principle can be central in characterizing intelligence in a broader sense (Wu 2020), assuming the availability of an effective representation for carrying out a predetermined task as an essential point for intelligent behavior.

The discussion considered so far requires the distribution $p(x, y)$ to be available. Whenever only a finite sample $\mathcal{D}$ is given, the information curve previously introduced exhibits a different behavior, as shown in Fig. 5.3. More precisely:

- If $I(X : T)$ is too small, *over-compression* comes into play, and a too much coarse representation compromises the accuracy $I(T : Y)$. In classical Statistical Learning terms, this situation would correspond to the underfitting case: a too-

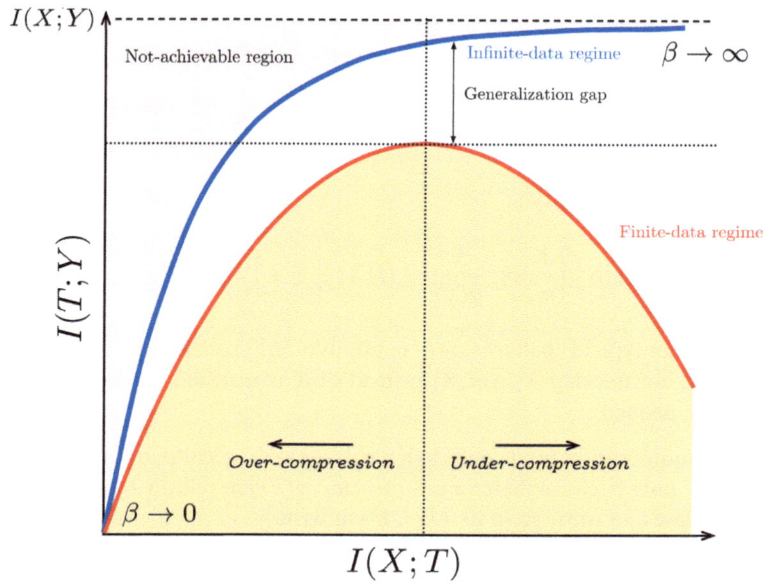

Fig. 5.3 *Information Bottleneck curve in the finite-data regime*, adapted from Tishby and Zaslavsky (2015). Only encoders living in the yellow area can be learned. The over-compression and under-compression regimes correspond—in the Statistical Learning language—to underfitting and overfitting respectively

high compression level for $\mathcal{T}$ produces the same effect as selecting a too-limited class of function $\mathcal{F}$ to fit $\mathcal{D}$ adequately.

- Conversely, if $I(X : T)$ is too near to $H[X]$, *under-compression* takes place, for which a too-detailed representation coexists with the sparsity of data, again compromising the accuracy level. In classical Statistical Learning terms—as when a too-large class $\mathcal{F}$ is selected—the overfitting problem occurs.

Between the two cases mentioned above, there exists an optimal level of compression for which the highest possible level of accuracy is achievable in the finite-data regime. The estimate (5.10) indicates how both dataset size m and compression level can intervene in controlling the generalization gap. Let us return to the learning problem by taking a Large-scale perspective, where the number of training examples and the input dimension are both huge. As highlighted by Tishby and Zaslavsky (2015), Shwartz-Ziv and Tishby (2017), and Shwartz-Ziv (2022), this approach provides some insights into explaining the functioning of Deep Learning protocols as procedures that provide compression in representation, shedding light on the overfitting problem and the benefits of adopting a layered structure.

Large-Scale Learning In this section we discuss a *typical-case bound* obtained by restricting the probability distributions to a suitable class. This result suggests a criterion for determining which problems can be effectively treated using an inductive protocol. Let us consider again the factorized case (5.5). Defining the set of ε-typical input of size m, i.e.

$$\mathcal{X}_{\text{eff}}(m, \varepsilon) := \left\{ x \in \mathcal{X} : \left| \frac{1}{m} \log \frac{1}{p(x)} - H[X_k] \right| < \varepsilon \right\}$$

the Chernoff-type inequality

$$\text{prob} \left(\left| \frac{1}{m} \log \frac{1}{p(x)} - H[X_k] \right| \geq \varepsilon \right) < e^{-2m\varepsilon^2}$$

ensures that nontypical patterns are negligible in probability for $m \rightarrow \infty$. Consequently, considering $\mathcal{X}_{\text{eff}}(m, \varepsilon)$ instead of $\mathcal{X}$ is justified. In this regard, some comments are needed.

- First, the equipartition holds only in a weak sense. Typical patterns are similar in probability only because their values of $-\log p(x)$ are within $2m\varepsilon$ of each other. As ε decreases, m must grow as $1/\varepsilon^2$. If we write

$$\varepsilon \sim \frac{1}{\sqrt{m}},$$

  then the most probable string in the typical set will be of order $2^{C\sqrt{m}}$ times greater than the least probable one, for some fixed C. On the other side, the typical set introduces a considerable simplification. Its elements have almost identical probability $2^{-mH[X_k]}$, and the whole set has a probability of almost 1. Consequently, we have roughly $2^{mH[X_k]}$ elements, and the other ones have no relevant role in probability.
- Second, the i.i.d assumption on X can be relaxed to *Markov field* structures. In this regard, let us assume that the probability measure P_X is factorized as

$$P_X = \prod_{i=1}^{m} P_{X_i|\text{Pa}_\text{i}},$$

  where Pa_i denotes the X-components adjacent to X_i, for which the following independence condition holds

$$X_i \perp\!\!\!\perp \neg\text{Pa}_\text{i}|\text{Pa}_\text{i}.$$

  If this Markov random field is ergodic, the Shannon-McMillan-Breiman theorem comes into play, and typical patterns approximate the entire patterns space $\mathcal{X}$ with high probability (Cover and Thomas 2012; Khinchin 1957). This *factorization hypothesis* seems reasonable in many application contexts—e.g. in image processing and speech recognition—where patterns comprise many local and weakly dependent patches.

While the main idea in the classical PAC framework is to exploit the concentration in probability around the Expected Error due to the Chernoff-bound effect, the typicality condition provides a new concentration in probability applicable to representations. Let T be a mapping of X, as in the learning apparatus (5.6). Ignoring the not-typical patterns, only $2^{H[X]}$ realizations of X can be considered. Moreover, on average, $2^{H[X|T]}$ typical realizations of X are mapped to the same value of T. Mimicking the classic argument to justify the Noisy Channel Coding theorem (MacKay 2003), the cardinality of the typical realizations of T can be estimated as

$$\frac{2^{H[X]}}{2^{H[X|T]}} = 2^{I(X:T)}.$$

Consequently, it is sufficient to consider the classification rules f defined for the representation of typical patterns, i.e. restricting ourselves to a proper subset $\mathcal{F}_T \subset 2^{\mathcal{X}}$. In other words, the substitution

$$|\mathcal{F}| \longmapsto |\mathcal{F}_T| = 2^{2^{I(X:T)}}$$

can be adopted in (5.3) obtaining a new lower-bound for the dataset cardinality to have the desired generalization control ε with probability $1 - \delta$:

$$n > \frac{2^{I(X:T)} + \log \frac{2}{\delta}}{2\varepsilon^2}. \tag{5.13}$$

At this level, three comments are needed.

- The bound (5.13) indicates that a more compressed representation exponentially reduces the amount of data needed to control the generalization gap via ε. However, it is necessary to keep in mind that, if $I(Y:T)$ is small, many possible representations with the same $I(X:T)$ are in principle available: the main point is to have an informative representation T on Y, and the estimate (5.13) is consequently significant when $I(Y:T)$ is sufficiently large.
- Reconsidering the expression (5.3) and introducing the set of effective functions $\mathcal{F}_T \subset 2^{\mathcal{X}}$ obtained via the representation T, the following relation follows

$$R[f] \leq \hat{R}_n[f] + O\left(\sqrt{\frac{2^{I(X:T)}}{n}}\right),$$

  becoming significant when $I(X:T)$ is smaller then $O(\log n)$, and in this case k bits of compression are equivalent to an exponential factor of 2^k training examples.
- By comparing (5.9) with (5.13), we immediately observe that the latter does not involve the $1/\varepsilon$ factor in the exponent. This fact is a direct consequence of the typicality assumption, informing us that the dataset size required to generalize

is significantly smaller. This conclusion aligns with our expectations: typicality provides us with an exponentially smaller space $\mathcal{X}_{\text{eff}}$ compared to eliminating negligible patterns—up to a threshold probability σ—within individual X_k and then considering the resulting m strings in $X_\sigma \times X_\sigma \times ... \times X_\sigma$.

5.4 Conclusions

While fully aware of the general scope of the generalization problem, we have explored the realm of Machine Learning to provide a precise formulation. The analysis proposed suggests the following general picture. By adopting the *Empirical Risk Minimization* principle, Machine Learning techniques can be understood in terms of *curve-fitting* procedures. Although certain quantitative guarantees regarding generalization are furnished when the hypothesis space is predetermined, inductive protocols that generalize well could respond to a partially different logic, by implementing *coarse-graining* procedures. The generalization gap control bounds discussed above contribute to justifying this approach, especially in the Large-Scale Learning regime. It remains to be clarified how a learning protocol concretely performs coarse-graining to construct an effective representation.

From our perspective, coarse-graining procedures are not only essential for establishing effective bounds but are also an inevitable step implicitly implemented by Deep Learning protocols. Even if the mechanisms of information compression—or the removal of irrelevant details—differ between artificial intelligence and biological intelligence, they still represent a significant point of connection. This analogy with coarse-graining in modeling could aid in both clarifying the limitations and potential of artificial intelligence and serve as a valuable conceptual tool within the framework of theoretical analysis. Although these aspects are not yet fully understood, we hope to have contributed to elucidating some of their fundamental premises.

Appendix 1: The Classical PAC Bound

Let us briefly review the classical *Probably Approximately Correct* (PAC) learning framework (Mohri et al. 2012; Shalev-Shwartz and Ben-David 2014; Valiant 1984). Fixed $f \in 2^{\mathcal{X}}$, the Expected Error $R[f]$ is a deterministic quantity. On the contrary, the Empirical Risk is random and depends on the sample realization $\mathcal{D}$. Thanks to the i.i.d assumption the identity

$$\mathbb{E}_{\mathcal{D}}[\hat{R}_n[f]] = R[f]$$

holds and $\hat{R}_n[f]$ converges to a Gaussian variable centered in $R[f]$ with a variance of order $O(1/m)$, as the Central Limit theorem prescribes. Moreover, the *Chernoff-bound* controls the probability of the Gaussian tails—i.e. the quantity to be minimized—so that

$$\text{prob}_{\mathcal{D}}(|R[f] - \hat{R}_n[f]| > \varepsilon) \le 2e^{-2n\varepsilon^2}. \tag{5.14}$$

At this point, relation (5.14) can not directly provide a generalization gap bound. Indeed, given the learning algorithm

$$\mathcal{A} : \mathcal{D} \longmapsto f_{\mathcal{D}} \in \mathcal{F},$$

the function $f_{\mathcal{D}}$ is a random variable dependent on the dataset $\mathcal{D}$. Therefore, a uniform control in $f \in \mathcal{F}$ is needed. In this regard, the *union-bound* trick is a standard tool in providing the following relations (Mohri et al. 2012):

$$\begin{aligned}
&\text{prob}_{\mathcal{D}}(\exists f^* \in \mathcal{F} : |R[f^*] - \hat{R}_n[f^*]| > \varepsilon) \\
&\le \text{prob}_{\mathcal{D}}\left(\bigcup_{f \in \mathcal{F}} \{f : |R[f] - \hat{R}_n[f]| > \varepsilon)\}\right) \\
&\le \sum_{f \in \mathcal{F}} \text{prob}_{\mathcal{D}}(|R[f] - \hat{R}_n[f]| > \varepsilon) \\
&\le 2|\mathcal{F}|e^{-2n\varepsilon^2}.
\end{aligned}$$

Consequently, given the confidence threshold $\delta := 2|\mathcal{F}|e^{-2n\varepsilon^2}$ with $\delta \in (0, 1)$, if the sample cardinality satisfies the inequality

$$n > \frac{\log|\mathcal{F}| + \log\frac{2}{\delta}}{2\varepsilon^2}, \tag{5.15}$$

the generalization gap is controlled by ε with probability $1 - \delta$ and uniformly in $f \in \mathcal{F}$. This time as f varies in $\mathcal{F}$, $\hat{R}_n[f]$ is expected to be a more irregular function than $R[f]$, the former being dependent on the specific dataset realization. In this regard, the uniform bound in f offers assurance about controlling the overfitting problem in minimizing $\hat{R}_n[f]$. More precisely, when the cardinality $|\mathcal{D}|$ is sufficiently large with respect to $\log|\mathcal{F}|$, the Empirical Risk becomes a good estimator for the Expected Error simultaneously for all $f \in \mathcal{F}$, as in (5.3). In other terms,

$$R[f] \le \hat{R}_n[f] + O\left(\sqrt{\frac{\log|\mathcal{F}|}{n}}\right) \tag{5.16}$$

and minimizing $\hat{R}_n[f]$ consequently becomes a good strategy to minimize $R[f]$. Moreover, $\log|\mathcal{F}|$ corresponds to the number of bits required to represent all elements in $\mathcal{F}$, thus serving as a natural measure of complexity when $|\mathcal{F}| < \infty$.

Appendix 2: The Covering Argument for $|\mathcal{F}| = \infty$

The union-bound trick provides a meaningful result only if $|\mathcal{F}| < \infty$ and the previous sample lower-bound is uninformative when dealing with an infinite function class. Generalizing (5.15) to $|\mathcal{F}| = \infty$ is possible by reducing the infinite case to the analysis of the finite one, adopting concepts as the *Rademacher complexity*, *grow function* or *VC-dimension* (Mohri et al. 2012; Shalev-Shwartz and Ben-David 2014). In what follows, we will discuss only the *covering argument* (Shalev-Shwartz and Ben-David 2014), although all these alternative approaches can be related to each other. As in the next paragraph, this perspective will also help determine a $\mathcal{F}$-independent bound. Let $\mathcal{F}$ be a compact space with respect to the topology induced by the metric

$$\Delta(f_1, f_2) := \text{prob}_p(f_1(x) \neq f_2(x)).$$

Therefore, the number of ε-spheres needed to cover $\mathcal{F}$—denoted by $N(\varepsilon)$—is finite and scales as $1/\varepsilon^d$, where d is the dimension of $\mathcal{F}$. The dimension d can be related to the concepts of VC-dimension, Hausdorff dimension, or topological dimension (Shalev-Shwartz and Ben-David 2014). In what follows, it will be thought of as related to the number of parameters needed to characterize an element f in $\mathcal{F}$. Moreover, for the binary classification problem, we have

$$R[f] = \text{prob}_p(f(x) \neq y)$$

and the following relation holds

$$|R[f_1] - R[f_2]| \leq \text{prob}_p(f_1(x) \neq f_2(x)) = \Delta(f_1, f_2). \tag{5.17}$$

In other words, if two functions belong to the same ε-sphere then the expected error of the former is approximated by the latter with an error at most ε. The same relationship is true also for the empirical risk by considering the empirical distribution $\hat{p}$ instead of p. At this point, given the center of each ε-sphere, the learning algorithm $\mathcal{A} : D \longmapsto f_{\mathcal{D}}$ can be replaced by an approximate version

$$\mathcal{A}_\varepsilon : D \longmapsto f_\varepsilon \in \mathcal{F}_\varepsilon,$$

where f_ε is the center of the ε-sphere in which $f_\mathcal{D}$ lives and $|\mathcal{F}_\varepsilon| < \infty$. Consequently, the substitution

$$|\mathcal{F}| \longrightarrow |\mathcal{F}_\varepsilon| := N(\varepsilon/2)$$

can be adopted in (5.15), obtaining a new lower-bound for n which depends linearly on d. Again, the main idea is the empirical risk concentration around the expected error via the Chernoff-bound. In this respect, the probability p comes into play only via the metric Δ. On the other side—once Δ is defined—the only object with a role is $\mathcal{F}$ via its dimensionality. Moreover, for a given n the generalization gap is uniformly controlled with probability $1 - \delta$ as follows

$$|R[f] - \hat{R}_n[f]| \leq \sqrt{\frac{d \log(2/\varepsilon) + \log \frac{2}{\delta}}{2n}}, \tag{5.18}$$

which is meaningless in the Deep Learning context, for which $d >> n$. This latter observation provides a relevant problem for Statistical Learning theory. Deep Learning models work with much fewer data than the classical bound would predict, also when the number of parameters exceeds the dataset size by orders of magnitude.

Appendix 3: $\mathcal{F}$-Independent Bound via Covering

The classical bounds (5.3), (5.15), (5.16) and (5.18) as discussed above require a class of function $\mathcal{F}$—or $\mathcal{F}_\varepsilon$—narrower than $2^{\mathcal{X}}$. On the contrary, if $\mathcal{F} = 2^{\mathcal{X}}$, the inequality (5.15) would require a number of data on the order of all possible realizations of X. In that case, the generalization gap control is meaningless. More in general, the classical PAC approach provides bounds that do not depend on the specific learning procedure adopted or sample distribution, thus characterized by the worst-case behavior. Appendix 2 offers a useful tool to rediscuss this point, with the difference that we assume $|\mathcal{X}| < \infty$ to obtain a bound independent of $\mathcal{F}$, and taking into account the properties of the probability distribution p. More in detail, a covering argument will be adopted by constructing an *effective partition* on $2^{\mathcal{X}}$ by identifying functions that agree on appropriate sets. Let us proceed step by step. Firstly, the following objects are defined as represented in Fig. 5.4:

$$\mathcal{X}_\sigma := \{x \in \mathcal{X} : p(x) > \sigma\} \quad \textit{with} \quad \sigma \in (0, 1),$$

$$\mathcal{F}_\sigma := \{f_\sigma : f \in 2^{\mathcal{X}}\},$$

$$f_\sigma(x) := \begin{cases} f(x) & \text{if } x \in \mathcal{X}_\sigma \\ 0 & \text{o.w.} \end{cases}.$$

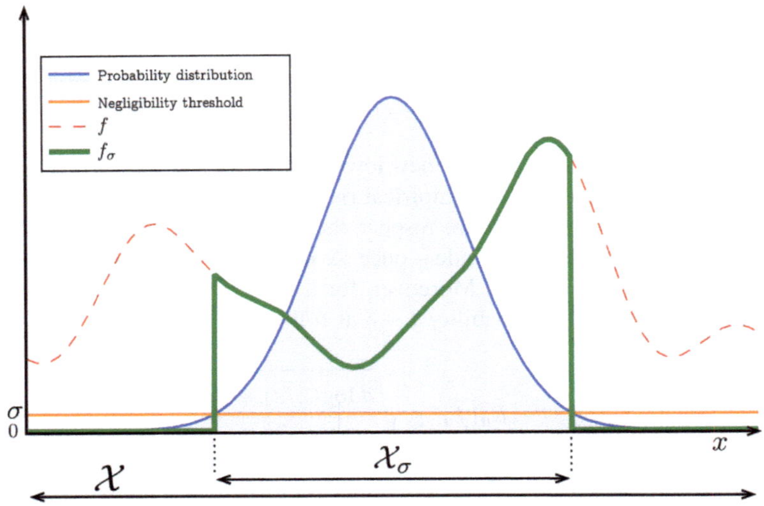

Fig. 5.4 *Non-negligible patterns $\mathcal{X}_\sigma$ and quotient on $\mathcal{F}$.* The procedure is schematically illustrated. For clarity, the above functions are continuous with $\mathbb{R}$-values

In other words, the set of *non-negligible patterns* $\mathcal{X}_\sigma$ is introduced by setting a threshold σ on the probability of each pattern. Moreover, the functions that agree on $\mathcal{X}_\sigma$ are identified by renouncing in characterizing them for negligible patterns: hopefully, this choice enables the adjustment of worst-case behavior, characteristic of the PAC approach. Secondly, the generalization gap is decomposed into three pieces as an immediate consequence of the triangular inequality and the union-bound trick:

$$\begin{aligned}\text{prob}_{\mathcal{D}}(|R[f] - \hat{R}_n[f]| > \varepsilon) \leq \text{prob}_{\mathcal{D}}\left(|R[f] - R[f_\sigma]| > \frac{\varepsilon}{3}\right)\\ + \text{prob}_{\mathcal{D}}\left(|R[f_\sigma] - \hat{R}_n[f_\sigma]| > \frac{\varepsilon}{3}\right)\\ + \text{prob}_{\mathcal{D}}\left(|\hat{R}_n[f_\sigma] - \hat{R}_n[f]| > \frac{\varepsilon}{3}\right)\end{aligned}$$

The three terms reported above can be separately considered. The general idea is to determine σ to set the first term to zero and simultaneously control the other two. Let us consider the first term by employing (5.17) and the *Markov inequality*:

$$\begin{aligned}|R[f] - R[f_\sigma]| &\leq \text{prob}_p(f(x) \neq f_\sigma(x)) = \text{prob}_p(x \notin \mathcal{X}_\sigma)\\ &= \text{prob}_p\left(-\log p(x) > \log \frac{1}{\sigma}\right)\\ &\leq \frac{\mathbb{E}_p[-\log p(x)]}{\log \frac{1}{\sigma}} = \frac{H[X]}{\log \frac{1}{\sigma}},\end{aligned}$$

where the Shannon Entropy $H[X]$ is assumed to be finite. Consequently, by defining

$$\frac{H[X]}{\log \frac{1}{\sigma}} := \varepsilon'$$

and choosing σ such that $\frac{\varepsilon}{3} \geq \varepsilon'$,

$$\text{prob}_{\mathcal{D}} \left(|R[f] - R[f_\sigma]| > \frac{\varepsilon}{3} \right) = 0.$$

The second term includes only $f_\sigma \in \mathcal{F}_\sigma$. Consequently, the classical bound holds. More precisely—observing that $|\mathcal{X}_\sigma| \leq 1/\sigma$ and adopting σ as above—the cardinality $|\mathcal{F}_\sigma|$ is controlled as follows

$$|\mathcal{F}_\sigma| = 2^{|\mathcal{X}_\sigma|} \leq 2^{\frac{1}{\sigma}} \leq 2^{2^{\frac{H[X]}{\varepsilon'}}}$$

and then

$$\begin{aligned}\text{prob}_{\mathcal{D}} \left(|R[f_\sigma] - \hat{R}_n[f_\sigma]| > \frac{\varepsilon}{3} \right) &\leq 2|\mathcal{F}_\sigma| e^{-2n\varepsilon'^2} \\ &\leq 2^{2^{\frac{H[X]}{\varepsilon'}}+1} e^{-2n\varepsilon'^2}.\end{aligned}$$

For the third term, the inequality (5.17) with the empirical distribution $\hat{p}$ allows us to use the standard Chernoff-bound again. More precisely—as already done for p—we have

$$|\hat{R}_n[f] - \hat{R}_n[f_\sigma]| \leq \text{prob}_{\hat{p}}(f(x) \neq f_\sigma(x)) = \text{prob}_{\hat{p}}(\hat{p}(x) < \sigma).$$

Consequently,

$$\begin{aligned}\text{prob}_{\mathcal{D}} \left(|\hat{R}_n[f_\sigma] - \hat{R}_n[f]| > \frac{\varepsilon}{3} \right) &\leq \text{prob}_{\mathcal{D}} \left(\text{prob}_{\hat{p}}(\hat{p}(x) < \sigma) > \frac{\varepsilon}{3} \right) \\ &\leq \text{prob}_{\mathcal{D}} \left(\text{prob}_{\hat{p}}(\hat{p}(x) < \sigma) - \mu > \frac{\varepsilon}{3} - \mu \right) \\ &\leq e^{-2n(\frac{\varepsilon}{3} - \mu)^2},\end{aligned}$$

where μ is defined as

$$\mu := \mathbb{E}_{\mathcal{D}}[\text{prob}_{\hat{p}}(\hat{p}(x) < \sigma)] = \text{prob}_p(x \notin \mathcal{X}_\sigma).$$

Remarkably, $\mu \leq \varepsilon' \leq \frac{\varepsilon}{3}$. Therefore, choosing σ such that $\varepsilon' = \varepsilon/6$ and putting it all together, the generalization gap is controlled by ε with probability $1 - \delta$ and uniformly in $f \in 2^X$ if n scales—unless a multiplicative constant—as

$$n > \frac{2^{\frac{6H[X]}{\varepsilon}} + \log \frac{2}{\delta}}{\varepsilon^2}.$$

Appendix 4: Elements of Information Theory

Let X be a random variable with discrete range $\mathcal{X}$ and probability distribution $p(x)$. The function $h(x) := -\log p(x)$ can be interpreted as a measure of the information content associated with $x \in \mathcal{X}$. Averaging over $\mathcal{X}$, the *Shannon entropy* $H[X]$ is then defined as

$$H[X] := -\sum_{x \in \mathcal{X}} p(x) \log p(x). \tag{5.19}$$

Entropy can be read as the minimum description length of X, i.e. the minimal number of bits needed—on average—to represent all the values of X. In this respect, $H[X]$ is a concave function in p that reaches its maximum $\log |\mathcal{X}|$ when p is uniform on $\mathcal{X}$, and the outcome of a random experiment is guaranteed to be the most informative as possible. Conversely, $H[X] = 0$ when X is deterministic. Moreover, (5.19) can be extended to the *joint* and *conditional* cases via

$$H[X, Y] := -\sum_{x \in \mathcal{X}, y \in \mathcal{Y}} p(x, y) \log p(x, y),$$

$$H[X|Y] := \sum_{y \in \mathcal{Y}} p(y) H[X|Y = y] := -\sum_{y \in \mathcal{Y}} p(y) \sum_{x \in \mathcal{X}} p(x|y) \log p(x|y),$$

which admit analogous interpretations. Conditioning allows us to isolate the information shared by X and Y. Given a particular input-output couple, a measure of how much information the output Y conveys about the input X is desired. In this respect, *mutual information* $I(X : Y)$ is defined as

$$I(X : Y) := H[X] - H[X|Y] = H[Y] - H[Y|X],$$

and can be interpreted as the mean information attributed to a realization of X given complete knowledge of Y, or vice versa. The equality $I(X : Y) = H[X] + H[Y] - H[X, Y]$ provides us with the more explicit formula that follows

$$I(X : Y) = \sum_{x \in \mathcal{X}, y \in \mathcal{Y}} p(x, y) \log \frac{p(x, y)}{p(x)p(y)}, \tag{5.20}$$

to be read as a special case of the *Kullback-Leibler divergence*. More precisely, suppose we have a priori a distribution p for X and consider the distribution q in place of p, given a certain set of experimental measures. Under the assumption that $q(x) = 0$ implies $p(x) = 0$, we can define the *KL divergence*

$$D[p||q] := \sum_{x \in X} p(x) \log \frac{p(x)}{q(x)},$$

to be interpreted as the information gap determined, on average, by using q instead of p. In the case in which $p(x) = q(x)$ for some x, then it will not contribute to $D[p||q]$. By employing *Jensen's inequality* for convex functions, *Gibbs' inequality* immediately follows (MacKay 2003):

$$D[p||q] \geq 0,$$

where the equality holds if and only if $p \equiv q$. As a corollary, $I(X:Y) = 0$ if and only if X and Y are independent. Moreover, p and q are similar when they have a low KL divergence, while a high KL divergence indicates dissimilarity. In this regard, *Pinsker's inequality* is a standard tool and implies that the KL divergence upper bounds the L^1-distance between distributions (Cover and Thomas 2012):

$$||p - q||_1^2 \leq \frac{1}{2 \ln 2} D[p||q]. \tag{5.21}$$

Consequently—although KL divergence is not a metric because it is not symmetric and the triangle inequality does not hold—it is still useful to think of $D[p||q]$ as a natural distance between p and q, as employed in (5.8).

References

Baldi, P. 2021. *Deep Learning in Science*. Cambridge: Cambridge University Press.
Bengio, Y., A. Courville, and P. Vincent. 2013. Representation Learning: A Review and New Perspectives. *IEEE Transactions on Pattern Analysis and Machine Intelligence* 35.
Bishop, C.M. 2007. *Pattern Recognition and Machine Learning (Information Science and Statistics)*. 1st ed. Berlin: Springer.
Cecconi, F., M. Cencini, and A. Vulpiani. 2013. *Chaos: From Simple Models to Complex Systems*. Singapore: World Scientific.
Chechik, G., A. Globerson, N. Tishby, and Y. Weiss. 2003. Information Bottleneck for Gaussian Variables. In *Advances in Neural Information Processing Systems*, ed. S. Thrun, L. Saul, and B. Schölkopf, vol. 16. Cambridge: MIT Press.
Cover, T.M., and J.A. Thomas. 2012. *Elements of Information Theory*. New York: Wiley.
Gilad-Bachrach, R., A. Navot, and N. Tishby. 2003. An Information Theoretic Tradeoff Between Complexity and Accuracy. In *Annual Conference Computational Learning Theory*.
Goldfeld, Z., and Y. Polyanskiy. 2020. The Information Bottleneck Problem and Its Applications in Machine Learning. *IEEE Journal on Selected Areas in Information Theory* 1: 19–38.

Goodfellow, I.J., Y. Bengio, and A. Courville. 2016. *Deep Learning*. Cambridge: MIT Press.
Harremoes, P., and N. Tishby. 2007. The information bottleneck revisited or how to choose a good distortion measure. In *2007 IEEE International Symposium on Information Theory*, 566–570.
Khinchin, A.I.A. 1957. *Mathematical Foundations of Information Theory*. Dover Books on Mathematics. Mineola: Dover Publications.
MacKay, D.J.C. 2003. *Information Theory, Inference and Learning Algorithms*. Cambridge: Cambridge University Press.
Mehta, P., C. Wang, A.G.R. Day, C. Richardson, M. Bukov, C.K. Fisher, and D.J. Schwab. 2019. A High-Bias, Low-Variance Introduction to Machine Learning for Physicists. *Physics Reports* 810: 1–124.
Mohri, M., A. Rostamizadeh, and A. Talwalkar. 2012. *Foundations of Machine Learning*. Adaptive Computation and Machine Learning Series. Cambridge: MIT Press.
Murphy, K.P. 2013. *Machine Learning. A Probabilistic Perspective*. Cambridge: MIT Press.
Ng, A. 2024. *Machine Learning Yearning*. Technical Strategy for AI Engineers, in the Era of Deep Learning. Edited by DeepLearning.
Schölkopf, B., F. Locatello, S. Bauer, N.R. Ke, N. Kalchbrenner, A. Goyal, and Y. Bengio. Towards Causal Representation Learning. *CoRR*.
Shalev-Shwartz, S., and S. Ben-David. 2014. *Understanding Machine Learning: From Theory to Algorithms*. Cambridge: Cambridge University Press.
Shamir, O., S. Sabato, and N. Tishby. 2010. Learning and Generalization with the Information Bottleneck. *Theoretical Computer Science* 411 (29): 2696–2711.
Shwartz-Ziv, R. 2022. Information Flow in Deep Neural Networks. Ph.D. Thesis.
Shwartz-Ziv, R., and N. Tishby. 2017. Opening the Black Box of Deep Neural Networks via Information. Published in arXiv.org.
Sidey-Gibbons, J., and C. Sidey-Gibbons. 2019. Machine Learning in Medicine: A Practical Introduction. *BMC Medical Research Methodology* 19.
Slonim, N. 2002. The Information Bottleneck: Theory and Applications. Ph.D. Thesis.
Slonim, N., and N. Tishby. 1999. Agglomerative Information Bottleneck. In *Advances in Neural Information Processing Systems*, ed. S. Solla, T. Leen, and K. Müller, vol. 12. Cambridge: MIT Press.
Soni, P., Y. Tewari, and D. Krishnan. 2022. Machine Learning Approaches in Stock Price Prediction: A Systematic Review. *Journal of Physics: Conference Series* 2161: 012065.
Tishby, N., and N. Zaslavsky. 2015. Deep Learning and the Information Bottleneck Principle. In *2015 IEEE Information Theory Workshop (ITW)*, 1–5.
Tishby, N., F. Pereira, and W. Bialek. 2001. The Information Bottleneck Method. *Proceedings of the 37th Allerton Conference on Communication, Control and Computation*, vol. 49.
Valiant, L.G. 1984. A Theory of the Learnable. In *Symposium on the Theory of Computing*.
Vapnik, V. 2013. *The Nature of Statistical Learning Theory*. Information Science and Statistics. New York: Springer.
Vapnik, V.N., and V.N. Vapnik. 1998. *Statistical Learning Theory*. A Wiley-Interscience Publication. New York: Wiley.
Wu, T. 2020. *Intelligence, Physics and Information – the Tradeoff Between Accuracy and Simplicity in Machine Learning*. Ph.D. Thesis.

Chapter 6
The Logic of DNA Identification

Sandy Zabell

Abstract The use of DNA identification evidence during the last several decades has revolutionized forensic science. But with the increasing complexity of the systems that are now being employed, a number of foundational challenges have come to the surface: the appropriate statistic to summarize the strength of the evidence, dealing with complex samples, searching data bases, safeguarding individual privacy, and effectively communicating results to a lay audience. In this survey, after describing the current system most commonly in use, I will present a typology of these issues, focusing on the recent advent of so-called probabilistic genotyping systems, which fit high-dimensional models, sometimes using Markov chain Monte Carlo estimation, and which can be technically challenging to explain to the trier of fact.

Keywords DNA identification evidence · Polymerase chain reaction · Short tandem repeat · DNA profile · Likelihood ratio · Cold hit · Odds ratio version of Bayes's theorem · Probabilistic genotyping system

6.1 Introduction

DNA identification evidence, developed by Sir Alec Jeffreys in 1985 and first used in a criminal case in 1986, has since become the gold-standard for forensic identification throughout the world.[1] There are several reasons responsible for its central role and importance in forensic identification: it is at once highly

[1] Its first use was in the arrest of the serial killer Colin Pitchfork in the United Kingdom, the subject of Joseph Wambaugh's 1989 book *The Blooding*.

S. Zabell (✉)
Department of Statistics and Data Science, Northwestern University, Evanston, IL, USA
e-mail: zabell@math.northwestern.edu

H. Hosni, J. Landes (eds.), *Perspectives on Logics for Data-driven Reasoning*,
Logic, Argumentation & Reasoning 35,
https://doi.org/10.1007/978-3-031-77892-6_6

discriminating, widely applicable, and ordinarily objective. This last aspect is in sharp contrast to many other forms of forensic identification evidence such as the use of bite marks, hair and fibers, and fingerprints, all of which are ultimately subjective and qualitative in nature.[2]

Today DNA identification evidence is used in many contexts, criminal, civil, and historical, including:

- Forensic cases (matching DNA evidence with either a victim or suspect)
- Paternity testing (identifying the father of a child)
- Searching DNA databases (for "cold hits" or relatives)
- Identifying bodies in mass disasters
- Missing person investigations
- Historical investigations

One celebrated example of the last is the identification of two missing children of Czar Nicholas II.

DNA identification evidence is inherently statistical in nature, and its logic depends on the context in which it is applied. But before discussing this logic, some biological and technological background is needed.

6.2 DNA 101

Our *chromosomes*, the repository of (most of) our DNA, are located in the cell nucleus.[3] The chromosomes come in pairs (one from the mother, one from the father), twenty-three pairs in all, and our genetic information is encoded in this DNA. Physically, DNA is a "double helix"—it has two intertwining strands. The genetic code consists of four "letters" A, T, G, C. That is, four different nitrogenous bases occur: adenine (A), guanine (G), cytosine (C), thymine (T), and each of the two strands in the DNA double helix consists of a sequence of these four letters. The two strands are "complementary"; that is, an A on one strand is paired with a T on the other, and a C on one strand pairs with a G on the other.

---

[2] In the case of fingerprints the classic example is the *Brandon Mayfield case*. Mayfield was a Portland, Oregon attorney who was arrested by the FBI in 2004 on the basis of a fingerprint found on a bag involved in the 2004 terrorist Madrid train bombings. Mayfield was held as a material witness for two weeks until the Spanish National Police (who had already expressed doubts to the FBI that Mayfield matched the evidentiary fingerprint) informed the FBI that they had identified the actual person who was the source of the fingerprint, resulting in Mayfield's release; see Zabell (2005). The Mayfield case illustrates a seed change in the world of forensic science: when DNA and fingerprints come into conflict—as they did in Mayfield—DNA wins.

[3] DNA also occurs in the mitochondria, as well as other forms of extrachromosomal DNA such as plasmids. Our discussion here, however, is confined to chromosomal DNA unless otherwise stated.

In sum: the information content in chromosomal DNA is captured by the sequence of the four letters in the genetic code found on either of the two complementary strands.

6.3 Forensic DNA 101

The current technology being employed in DNA identification uses PCR: the *polymerase chain reaction*. This involves a form of "molecular xeroxing", in which after the DNA is extracted from cells (most commonly blood, semen, or saliva) the number of copies of the DNA is increased manyfold. [4]

DNA identification is based on the quest for *hypervariable polymorphisms*; the more the DNA in a given region varies from person to person, the more discriminating it will be. In the current systems, these involve STRs (*short tandem repeats*). For reasons that are not understood, throughout the genome there are locations where a sequence of bases is repeated several times, for example

$$ATCC\ ATCC\ ATCC \ldots ATCC \qquad n \text{ times}$$

Such short tandem repeats occur in noncoding regions of the DNA; the resulting absence of selective pressure in these is what permits them to be hypervariable.[5] Determining the number of repeats for a particular region (or "locus") is done by performing *capillary electrophoresis*: the longer STRs migrate more slowly in an electrical field. The result is an *electropherogram* (which, roughly speaking, resembles an EKG) with a sequence of peaks from which one can read off the

[4] Imagine if you will a xerox machine that can make only one copy of a document, but one of arbitrary length. If you take a one-page document and copy it, you will now have two copies. If you then insert these two pages back into the machine, treating them as a single two-page document, both are copied and you now have four; insert these four in turn and you have eight, *und so weiter*. If you repeat this procedure n times, then you end up with 2^n copies.

The polymerase chain reaction is a complex biochemical process that accomplishes this on the molecular level by repeating a series of three steps: first, separating the paired strands of the DNA double helix by raising the temperature of the solution to approximately $95°C$ ("denaturation"); then lowering the temperature to approximately $55°C$, thereby permitting fluorescently tagged "primers" (short segments of DNA complementary to a region of the DNA next to the targeted region) to attach to this region ("annealing"); and then finally raising the temperature to an intermediate level (approximately $75°C$), permitting the enzyme DNA polymerase to extend the primer, thus generating the rest of the complementary strand ("extension"). The process is termed "thermo-cycling", each cycle resulting in a doubling of the DNA. Commonly thirty cycles are performed, resulting therefore in (assuming perfect efficiency) a billion-fold increase in the DNA. This has the merit of increasing the sensitivity and applicability of the test, but also the potential for detecting low level contamination and trace amounts of DNA.

[5] Most mutations in genes coding for proteins are deleterious and so are selected against. One example is sickle cell anemia, which is the result of a mutation in the gene coding for hemoglobin.

Table 6.1 The nine Profiler Plus loci

Blue	D3S1358	vWA	FGA
Green	D8S1179	D21S11	D18S51
Yellow	D5S818	D7S820	D13S317

number of repeats at each locus. As result, each STR produces a pair of repeat numbers (since our chromosomes come in pairs), the *genotype* for that locus.

The result is a *DNA profile* typically consisting of 9–15 pairs of repeat numbers (although there are some commercially available test kits which test for more). For example, in one kit there are nine "Profiler Plus" loci, denoted "Blue", "Green", and "Yellow" (three each, see Table 6.1) and four additional "Cofiler" ones. (The blue, green, and yellow refer to the fluorescent dyes attached to the primers, permitting the fragments to be detected during electrophoresis.)

Note that one does *not* look at all the DNA, but only a very small set of locations comprising only a very small fraction of the genome.

Putting this all together, there are several steps, technical and analytical, in DNA typing:

- extraction: lysing the cells (to get DNA)
- amplification: focussed on a small number of targeted sites (via PCR)
- capillary electrophoresis: separating STRs by length via migration speed
- genetic typing: determining the DNA profile by analyzing the electropherogram generated from the last step.

6.3.1 Computing Genotype Frequencies

The first step in computing a DNA profile's rarity is to determine the allele frequencies. This is in principle straightforward: the allele frequencies are determined by sampling from different human populations. Such data appear in publications such as the *Journal of Forensic Sciences*, and vary among different racial and ethnic groups. For example, for locus D3S1358, an FBI database indicates that among 203 Caucasian individuals having a total of 406 alleles, the 16 allele occurs 94 times, so the estimated frequency is 94/406, or 23%, and the 18 allele occurs 66 times, so the estimated frequency is 66/406, or 16%.

Alleles come in pairs at each locus, one allele contributed by the father, one allele by the mother. So for D3S1358, what is the frequency of a 16, 18?

In general, if alleles A, B have frequencies p_A, p_B, and *independence* holds, then the frequency of the AB genotype is

$$p_A p_B + p_B p_A = 2p_A p_B$$

(in this case the population is said to be in "Hardy-Weinberg equilibrium" at this locus). So the answer is $2(0.23)(0.16) = 0.07$, or about 7%.

Note that in this example AB is a *heterozygote*; for the *homozygote* AA, one uses the formula p_A^2 instead.

Ordinarily one examines multiple loci, and the totality of genotypes at the typed loci is termed an individual's *DNA profile*. The different loci are selected to be (at least approximately) statistically independent (this is termed "linkage equilibrium"), and so the overall frequency of the detected profile is computed by simply multiplying the different genotype frequencies (the "product rule"). (Small adjustments are also available if more accurate estimates are desired.)

6.4 The Unifying Framework of the Likelihood Ratio

In forensic DNA identification, one typically considers a pair of competing hypotheses, say H_0 and H_1, and some form of DNA evidence E relevant to them. If $P(E \mid H_k)$ is the conditional probability of E assuming H_k, it is generally agreed that the relevant quantity in such situations is the *likelihood ratio* (LR), here $P(E \mid H_1)/P(E \mid H_0)$, comparing the probability of seeing E under the two competing explanations. The value of the LR depends on the nature of the propositions in question, providing both a common framework in which to view the evidence, as well guidance on how to quantitatively express its value.

In the following we will consider several different scenarios and the resulting different LRs.

6.4.1 Source Attribution

Most commonly there is an evidentiary sample, consisting of blood, semen, or saliva, say, and the question is its source (usually a victim or suspect). Suppose, for example, someone has been sexually assaulted, semen has been recovered on a vaginal swab, and at a locus both the evidence and suspect have the same type, AB. This concordance certainly supports the prosecution contention that the suspect is the rapist, but what weight do we assign to this evidence? We can represent this situation as:

$$\begin{array}{ccc} \text{Evidence} & & \text{Possible source} \\ AB & \overset{?}{\longleftarrow} & AB \end{array}$$

The two hypotheses are

H_0: Random individual (someone other than the suspect) is the source

H_1: Suspect is the source

and the likelihood ratio in this case is (to a first approximation)

$$\frac{P(E \mid H_1)}{P(E \mid H_0)} = \frac{1}{2p_A p_B}.$$

That is, if the suspect is the source, then the probability of seeing the DNA evidence AB is one, while if the prosecution is wrong, and some other, "random" individual is the source, then the probability of seeing the AB is just the frequency of the AB in the population, given by the Hardy-Weinberg formula discussed earlier. That is, the LR is just the reciprocal of the frequency of the genotype in the population: the rarer the type, the stronger the evidence that the suspect is indeed the source of the biological evidence.

6.4.1.1 Refinements

There are a few refinements worth noting here. First, the frequency of STR genotypes varies among populations, and so frequencies are often given for a range of populations (in the US, usually African American, Caucasian, and Hispanic), or a weighted average of these. The so-called "ceiling principle" was an approach advocated by the first NRC committee to deal with this issue: one took the maximum frequency of an allele over all populations for which data was available for the allele in question. Although "conservative" (in the sense that it gave an upper bound for the observed profile frequency), it was criticized by a number of population geneticists on the grounds that it was unscientific, because it did not represent the actual frequency of the profile in any real population, and might in fact deviate from any actual profile frequency by several orders of magnitude.

One fallacy sometimes seen is to take the race or ethnicity of the *suspect* (or, more generally suspected source, perpetrator or victim) as representing the appropriate reference population. This is in fact exactly the wrong way round: the race or ethnicity of the suspect is irrelevant since their profile—rare or common—is already a given, and the relevant question is instead the frequency of the profile in the population of potential contributors. One extreme example here might be a situation where the possible sources of the evidence are thought to be members of a very special subgroup (for example, if a Federal agent found murdered on a Native American reservation had blood on them thought to come from the assailant). This can be challenging if frequency data is unavailable for that subgroup.

Another issue is the possibility of some kind of error in the typing of either the suspect or the evidence. (This could happen if, for example, samples are accidentally switched, or some form of contamination occurs.) If p is the overall frequency of the DNA profile over all loci, α the false negative rate, and β the false positive rate, then

$$\frac{P(\text{match}|\text{suspect is source})}{P(\text{match}|\text{chance})} = \frac{1-\alpha}{(1-\alpha)p + \beta(1-p)}.$$

If p is very small (i.e., much smaller than either α or β, as is typically the case), then this is approximately $(1-\alpha)/\beta$; and in any case is bounded from above by $1/\beta$. That is, the strength of the evidence is limited by the chance of an erroneous inclusion, be it due to clerical error, accidental switching of samples, or whatever.

This is important because although it is difficult to assign a precise numerical value to the probabilities of such errors, it can scarcely be argued that they are anywhere near 10^{-10}. Unfortunately, the siren-like attraction of supposedly astronomically small probabilities can often blind one to their practical limits. In the end the value of forensic or any other type of evidence is totally dependent on the reliability and validity of the process by which it is generated. (This point is discussed later on in the paper.)

6.4.2 *Paternity*

In paternity cases the type of the mother and child are known, and the issue is whether an accused man is the father. Consider the following situation:

Known mother: A, B	Known child of mother: A, C
Putative father: C, D	C is the *paternal obligate allele*

(The evidence)
H_0: Random individual is the father
H_1: Putative father is the true father
The likelihood ratio is

$$\frac{P(E \mid H_1)}{P(E \mid H_0)} = \frac{\frac{1}{2}}{p_C} = \frac{1}{2p_C}.$$

The reasoning is that the mother must contribute the A, so the C must come from the father; it is the "paternal obligate allele". Under H_0 (the father is a random unknown individual), the probability that the allele he contributes is a C is just the frequency of C in the population; that is, p_C. Under H_1 (the putative father is indeed the true father), the probability he contribute his C rather than D is $1/2$. (Note we are conditioning here on the genotypes of all three, and the fact that the child's mother is known.)

Obviously the specific formula will depend on the genotypes of mother, child, and questioned father; it is an instructive exercise to work out what the LR is in the other cases (for example, mother and child are heterozygous, but the putative father is homozygous).

Even in this apparently simple setting, there are a few interesting phenomena worthy of note.

6.4.2.1 An Apparent Paradox

Note that even when the putative father is consistent, the LR can *decrease*! This is because we can have

$$p_C > \frac{1}{2} \implies 2p_C > 1 \implies \frac{1}{2p_C} < 1.$$

Admittedly this sort of thing does not occur in practice, since not surprisingly the STRs employed in DNA testing do not have alleles with a frequency of greater than 1/2, but it does illustrate that one's untutored intuition in these cases can be treacherous, hence the utility of the LR in quantifying the genuine strength of the evidence.

6.4.2.2 Neither Explanation May Be Likely

Although in general *there is no partial credit in DNA* (that is, if you are excluded at one locus, you are excluded, period), this is not the case for paternity attribution because *mutation* is possible: the allele contributed by the father may have mutated and then multiplies during development. (Here the paternal allele comes from a single cell, while in the source attribution case the evidence consists of many cells, and the fact that one or just a few of them may have mutated is of no consequence.)

Suppose for example that there are 10 consistent loci and 2 inconsistent loci. Then $P(E \mid H_0)$ and $P(E \mid H_1)$ are *both small.* This is because on the one hand a random individual is unlikely to match up at as many as ten loci; if the, say, probability of a match is 1 in 10 (a reasonable order of magnitude in these STR systems), and X is the number of matching loci, then X has a binomial distribution with parameters $n = 12$ and $p = 1/10$, and the binomial probability $P(X \geq 10)$ is less than one chance in 180 million. On the other hand if the putative father is the true father, then the alleles he contributes are unlikely to have undergone two mutations, both because mutations in general are relatively rare, and because in particular the STR loci used in paternity testing are chosen to have especially low mutation rates. So in this case *the LR is a balance of improbabilities*, but it helps us to determine which explanation is to be preferred and by how much.

Many laboratories handle this in an ad hoc fashion. For example, the laboratory policy might be that if there is just a single inconsistency this is still taken to be strong evidence the putative father is the true father, while two inconsistencies is taken to be inconclusive, and three inconsistencies an outright exclusion. Such a rule might make sense as a qualitative summary if it is consistent with the computed LRs.[6]

---

[6] A referee has suggested this might be another case where a handling or contamination error, ordinarily a small probability event, could become relevant. Such handling errors can certainly happen. In one case this author is acquainted with, a mother accused a man of being the father of her child, but DNA testing excluded the putative father. After the mother strenuously objected to the test results on the grounds that there were no other possible candidates, the testing laboratory became uncomfortable and initiated an internal audit. They subsequently discovered that the supposed DNA profile of the putative father was in fact the DNA profile of a different man whom they had earlier tested.

6.4.3 Body Identification

A body is found thought to be the missing child of two parents. The important logical difference between this and the paternity case is that now the types of the father and mother are known, and the parentage of the victim is what is in question. Suppose the given data are:

Known mother: A, B Known father: C, D

Body: A, C

(The evidence)

H_0: Body is a random individual

H_1: Body is that of the missing child

In this case the likelihood ratio is

$$\frac{P(E \mid H_1)}{P(E \mid H_0)} = \frac{\frac{1}{4}}{2p_A p_C} = \frac{1}{8p_A p_C}.$$

(Once again, of course, the appropriate formula depends on the genotypes of the mother, father, and child—whether heterozygous or homozygous—as well as the shared alleles of the three. The reader is encouraged to work this out for several possible scenarios.)

6.4.4 "Cold Hits": "Probable Cause" vs. Cold Hit Scenarios

So far we have not discussed why someone became "a person of interest", at least to the extent that we have bothered to determine their DNA profile. Sometimes we have a considerable amount of prior information pointing to a specific individual (the *probable cause* scenario), but in other cases we may have little or no prior information pointing to a specific person. In such cases law enforcement may turn to searching a forensic database of individuals and may, if they are lucky, find a matching profile (the *cold hit* scenario).

How does one assess the strength of the evidence in such situations? A common intuition is that the evidence is more compelling in the probable cause setting than the case of a cold hit, where someone may be identified by trawling through a database of potentially millions of individuals. In fact this has been the subject of considerable debate and distinguished statisticians have differed on what to do.

Forensic DNA identification evidence has been the subject of two reports of the US National Research Council, and one of the issues the two reports differed on was this one. Interestingly, this is a situation where one's statistical philosophy (objective and physical vs. subjective and Bayesian) can play an important role. The first report (NRC1 1992) effectively skirted the issue by recommending that when searching a database you use some of the test loci to do the search, and—if there is a hit—you use the other loci to compute the likelihood ratio for the matching profile. The thought was that this way the loci used in quantifying the weight of the evidence of the match had not been tainted—if you will—by their use in the search process since they had not been used at that stage.

It would be a serious understatement to say this proposal did not meet with widespread approval by *either* of the two warring camps. Those who thought the weight of the evidence was effectively the same in the two scenarios thought this would just be an unnecessary waste of precious evidence, while those who thought the evidentiary status of the DNA evidence must surely be different (and weaker) in the cold hit case regarded this ad hoc approach to be merely an evasion, avoiding rather than confronting the issue.

The 1990s were a period of rapid change in forensic DNA identification evidence (most notably in the shift from the older RFLP technology to the new wave STR one). In 1996 the NRC revisited the subject, convening a second panel which issued a new report (NRC2 1996). This new panel agreed that in the case of a database search some adjustment was called for, and came up with a novel proposal: if the cold hit was the result of searching a database of size n and p was the standard match probability, then the recommendation was to use $1 - (1 - p)^n \approx np$ instead of p. The reasoning behind this was simple: if p is regarded as the probability that a random profile in the database would match the evidentiary profile, then $1 - p$ would be the probability of a non-match, $(1 - p)^n$ the probability that *none* of the profiles in the database will match (assuming independence), and therefore $1 - (1 - p)^n$ the complementary probability of seeing at least one match. Thus the question being asked and answered is: how unusual is it to see a chance match in the database? If this probability is sufficiently small, then this casts doubt on the hypothesis that the observed match was purely fortuitous, and therefore points instead to the alternative hypothesis that the source of the evidentiary DNA was in fact present in the database. This is effectively a classical test of statistical significance, the outcome being to incriminate the database.

This interesting proposal also failed to gain widespread support. If you thought the original match probability p was the right number, then if you searched a database of a million people the result would be to cut down the database match probability by six orders of magnitude (the result of multiplying by a million), while those who thought the match probability was different did not find the underlying logic convincing. The point is that the relevant quantity is not so much the size of the database but that of the potential population of suspects. To see the nature of the problem, suppose that $n = 1000$, $p = 1/10,000$, and that in the mid-size city in which the crime took place, there are $N = 1,000,000$ potential suspects. Then the NRC2 rule would replace p by $np = 1/10$ (so that the evidence is much weaker but still points in the right direction), but if p were applicable to the population of suspects, then one would expect on the order of 100 individuals with matching profiles, of which the database has enabled us to identify one (so that the probability that they are the actual source would be only about 1 in 100).

But the resolution of this (apparent) paradox is really quite simple using the *odds ratio version of Bayes's theorem*:

$$\frac{P(H_1 \mid E)}{P(H_0 \mid E)} = \frac{P(E \mid H_1)}{P(E \mid H_0)} \cdot \frac{P(H_1)}{P(H_0)};$$

where

- E is the DNA evidence
- H_0 the hypothesis that the target is not the source of the DNA
- H_1 the hypothesis that the target is the source of the DNA

In words: the posterior odds in favor of one hypothesis versus the other is the product of the prior odds times the likelihood ratio; the likelihood ratio incorporates the DNA evidence as the multiplicative statistical factor transforming the prior odds (prior, that is, to the DNA evidence) into the posterior odds after learning the DNA evidence.

So it turns out *both* sides were right. The impact of which scenario we are in (probable cause vs. database search) does not significantly affect the likelihood ratio (see below), but the prior odds in the two scenarios clearly differ: in the probable cause case there is ordinarily some prior substantial evidence pointing towards the target as the source (there is presumably a reason their DNA was typed in the first place), while if you search a database this is typically because you have no viable candidates in view (that is why you opted to search the database).

Why is the likelihood ratio largely unchanged? This is essentially an instance of what is termed the likelihood principle: the likelihood ratio does not depend on how the evidence was acquired; the evidence is the evidence. (So for someone who subscribes to the likelihood principle, when tossing a coin repeatedly to determine its probability of coming up heads, there is no difference between tossing a coin 10 times and seeing nine heads and one tail in any order, vs. tossing a coin until the first tail comes up, which turns out to be on the tenth toss.)

But why the qualification "largely" unchanged? There is in fact a clear difference between the two scenarios in terms of the DNA evidence: in one case you have only matched the DNA profiles of the target and the biological evidence, in the other you have not only matched the evidence to a sample in the database but have also *excluded every other member of the database as a potential source*. (Obviously finding multiple matches in the database raises a whole set of other issues, but is quite unusual nowadays given the current discriminating power of DNA identification methods.)

To understand this last point better, consider a crime committed in a secure prison facility where one knows the identity of all the individuals who might be responsible. (This type of situation is termed the "island problem".) If there is a total of n such individuals, inmates and staff, all of whom have been tested and precisely one of whom matches the evidence, then obviously we have identified the source of the evidence given the hypothetical. But suppose instead that in addition to the inmates and staff there are other people who had access to the facility at the time in question, but who are not readily available for testing. (For example, visitors, outside one-time contractors, touring legislators, etc.) If there are N such individuals, then our testing has reduced the population of possible suspects from an initial $n+N$ to $1+N$, one of whom is known to match. The issue of the quantitative evidentiary impact of the mass testing then reduces to a question of the relative sizes of n and N.

If at one extreme $N = 0$, then, as we have just seen, the source has been definitively identified (at least up to the validity of our assumptions). If instead $n = 1$, then this exactly the evidentiary situation we find ourselves in when discussing the probable cause scenario (albeit typically with a large value of N). It is intuitively clear that the larger the value of N, the less the value of the posterior odds after; or equivalently, the smaller the value of N, the greater the value of the posterior odds, and this decrease or increase is necessarily reflected in the change in the likelihood ratio since the prior odds have not changed. It is an instructive exercise to work this out quantitatively under a reasonable set of assumptions.

Explaining these issues to the trier of fact can be complicated (especially since it is not the role of the DNA analyst to provide prior odds to them).

6.4.5 Familial Searches

One variation on the theme of a database search is to search a database for a "near miss". This is because relatives are much more likely to have similar profiles than a randomly selected member of a population. For example, in one state case the FBI was sent a semen sample in a rape case, and found that of four heterozygous loci tested (so eight alleles in total), seven out of the eight alleles detected matched those of the suspect. (This was back in the RFLP days, when fewer but more discriminating loci were examined.) *But there is no partial credit in forensic DNA*: an exclusion is an exclusion is an exclusion (even if at just one locus). And so the FBI quite rightly reported back to the state laboratory that their suspect could not have been the perpetrator of the crime. But fortuitously matching seven out of eight alleles purely by chance is quite unlikely, and so the FBI also noted in their report that state law enforcement might look into close relatives of the (now exonerated) suspect. And indeed, one of the brothers of the suspect turned out to match all eight alleles that had been detected (and was subsequently charged and convicted).

Nowadays familial searches are not uncommon and there are some spectacular success stories associated with their use (for the case of the "Golden State killer", see Phillips, 2018). Of course there are privacy and legal concerns as well; see, e. g., Kaye (2013).

6.4.6 *Mixtures*

One important complication in all this is that often two or more sources of DNA are present. For example, in a rape case an intimate swab taken from the victim will contain both the victim's and the rapist's DNA, and DNA from the victim's husband or boyfriend might be present. There are a variety of ways to analyze such samples, ranging from the (relatively) simple to the highly complex. For example, one might see n alleles $A_1, \ldots, A_n$ in an evidence sample. There are

$$n + \binom{n}{2} = n + \frac{n(n-1)}{2} = \frac{n(n+1)}{2}$$

consistent genotypes (n homozygotes and $n(n-1)/2$ heterozygotes), a quantity that grows rapidly with n.

6.4.6.1 The CPI

One simple approach to analyzing this is the CPI (*Combined Probability of Inclusion*), which uses

$$(p_{A_1} + \cdots + p_{A_n})^2.$$

(Intuitively, think of the n alleles as being one "super-allele", with the CPI being the probability of seeing a homozygote.)

The CPI (sometimes referred to in the paternity setting as the probability of a *random man not excluded*) has the virtue of simplicity and being relatively easy to explain (the chance that a randomly selected person would not be excluded as a potential contributor to the mixture). (The use of the terms "included" and "excluded" here are themselves the subject of debate as being misleading to the layperson.)

6.4.6.2 The Likelihood Ratio

The CPI also has the disadvantage of not using all the information available, and most forensic scientists nowadays favor the use of the likelihood ratio. This was in fact recommended by NRC2, but was initially much less commonly used.

For example, suppose four alleles seen at a locus: A, B, C, D, and that the two competing hypotheses are:

H_0: Two unknown random individuals contributed to the mixture.

H_1: Unknown individual A, B + suspect C, D contributed to the mixture.

In this case the likelihood ratio is (see NRC2 1996, p. 130 and Table 5.1 on p. 163):

$$\frac{P(E \mid H_1)}{P(E \mid H_0)} = \frac{2p_A p_B}{24 p_A p_B p_C p_D} = \frac{1}{12 p_C p_D}.$$

More generally, ***if*** all alleles are present and accounted for, a relatively simple general formula for the likelihood ratio exists; see Weir et al. (1997). (The formula and its derivation are a simple application of the inclusion-exclusion method.)

If .... (as the Spartans once told a Persian envoy). In fact many complications may exist:

- The number of contributors may be unknown
- Different amounts of DNA may be contributed by different individuals
- "heterozygote imbalance" (the two peaks are of different height)
- "stutter" (there are peaks present not corresponding to an allele)
- "Allelic dropout" (the absence of alleles expected to be present)

Let us briefly consider each of these in turn.

First, there is the number of contributors. There are certainly situations where this is unknown. For example, one may be dealing with a nasty case of sexual assault where there are multiple assailants but the exact number is unknown. This is a problem because the standard likelihood ratio approach assumes the number of contributors c is known (as opposed to the CPI). A natural move here is to compute the LR assuming different values for c, but of course these may give very different qualitative results. Statistical methods are being developed to estimate c, but this is very much a work in progress.

Next, there is the issue of differing amounts of DNA present from each of the contributors to the mixture. (If some of these stand out much more than others, it is common to refer to "major" and "minor" contributors to the mixture.) This can complicate the interpretation of the electropherogram: a major contributor can "mask" a minor, as well as how to assess heterozygote imbalance and stutter.

Heterozygote imbalance If an individual is heterozygous at a locus, then one will see two peaks on the electropherogram corresponding to the two alleles. If the process of PCR amplification were exact, then the two alleles would be present in exactly the same quantity, and the two peaks would have exactly the same height. But life and PCR is imperfect, and there will always be some difference in the height of the two peaks, reflecting statistical variations in the different stages of the PCR process. As a result, in the dark ages of DNA interpretation an analyst would employ rules of thumb to judge whether the difference in heights was so great as to mean the two alleles must necessarily have come from two different contributors, or instead were close enough in height to be consistent with (but not necessarily!) coming from the same contributor. An example of such a rule of thumb is that the shorter of the two peaks be within 80% the height of the taller.

Stutter The polymerase chain reaction process is an impressive technological achievement, winning Kary Mullis (1924–2019), its inventor, the Nobel Prize in Chemistry in 1993. But its very complexity (for example, the choice of which STRs to use, the different temperatures in the thermo-cycle, the length of time each step takes in the cycle, optimizing all these parameters to permit amplifying all STRs simultaneously) means it is necessarily imperfect, and as a result can generate a number of artifacts. The most common of these is stutter.

Stutter is the phenomenon that due to imperfections in the amplification process, one sometimes sees a small peak to the left of a true peak, representing amplicons one repeat shorter than the ones producing the main peak. The copying of the DNA strand during extension has prematurely terminated, and the result is a fragment shorter by one than it should be. (Other kinds of stutter are possible but less common.) This gives rise to the question: does a short peak in stutter position represent an actual allele or not? Here too there are rules of thumb, such as a stutter peak can be no higher than 10%, say, although this percentage can depend on which STR locus is being considered.

This opens the door to all sorts of subjectivity. Suppose a potential source matches its target profile at all loci but one, where there is a peak in stutter position whose height is 20% of the main peak, even though the posited source does not have an allele corresponding to this position. It would be hard to imagine an analyst declaring this an exclusion. Or supposing a minor contributor, present at only 10% that of the major, does have an allele with a corresponding peak in stutter position in the evidence, then an analyst would surely declare this peak in stutter position to represent a "righteous allele", even though they would only score it as stutter if the minor did not have an allele with an expected peak in this position.

Finally, there is *allelic dropout*. Precisely because PCR is so sensitive and capable of detecting minute quantities of DNA, DNA analysts often work with such minute quantities of DNA that not all the alleles in the evidence are detected. They have "dropped out". Here too it is human nature to explain away clear discrepancies in evidence such as missing alleles as due to dropout. (One common fallacy here is to use the CPI to interpret a mixture even when some of the alleles in an hypothesized source are not present in the evidence. In effect, one is calculating the probability that the alleles of a random person would be a subset of those in the evidence sample, even though some of the alleles of that person are actually missing in the sample!)

In forensic science *one is on the road to perdition if the interpretation of evidence is influenced by knowledge ahead of time of the source one is trying to match it to.*

6.4.6.3 Technical Issues

For more than 15 years the scoring of alleles was done by hand, the analyst reading the peaks on the electropherogram, employing

- "stochastic thresholds" (below which peaks are not used for statistics);
- "analytical thresholds" (below which peaks cannot not be scored);
- "peak height ratios" (rules of thumb for when peaks can be paired)
- "stutter percentages" (rules for when a peak can be scored as stutter)

The distinction between stochastic and analytical thresholds is an interesting one. Analytical thresholds are familiar from chemistry, and in this setting refer to drawing a line below which any observed signal is regarded as noise and therefore is disregarded. On the other hand, stochastic thresholds (which are greater than analytical ones) refer heights above which a peak can be used for purpose of statistical frequency calculations. This leaves a demilitarized zone, between the analytical and stochastic thresholds, where peaks can be used for purposes of qualitative interpretation but cannot be used quantitatively in determining a profile frequency.

We have already encountered peak height ratios and stutter percentages. Far from being a straightforward process, this can involve a considerable element of subjective interpretation, lab protocols ordinarily leaving the analyst great leeway in scoring alleles (at least until recently).

A disturbing example of what could happen was reported in a paper in the *New Scientist* (2014). This involved asking 17 analysts to re-interpret the electropherogram of a complex mixture that they had previously analyzed (as an *inclusion*). The results revealed a disturbing lack of consistency among the analysts:

- 1 analyst agreed with their previous finding;
- 4 analysts now judged the evidence inconclusive;
- 12 now interpreted the evidence as an exclusion.

The cause was not hard to identify: as John Butler (then at the US National Institute of Standards and Technology, NIST) noted (Butler 2015),

> Many labs are doing or attempting more complex mixtures often without appropriate underlying validation support or consideration of complicating factors.

Recognition of this problem has led to the development of much more sophisticated and objective approaches to the interpretation of mixtures, culminating in the current probabilistic genotyping systems.

6.5 A Hierarchy of Interpretive Models

Current approaches that have been developed can be classified into a hierarchy of increasingly more complex methodologies:

1. binary (classical)
2. semi-continuous (modern)
3. fully continuous (postmodern)

A PGS (*probabilistic genotyping system*) employs (2) or (3).

6.5.1 Binary Models

This is the classical approach to complex mixtures noted earlier. In the basic (qualitative) binary model, alleles are scored as present or absent, and ideally one then uses the Weir et al. (1997) and Curran et al. (1999) formulas to compute the likelihood ratio, summing probabilities of consistent genotype sets, aided by suitable software such as POPSTAT or DNAMIX I, II.

Such an approach is inappropriate if dropout has occurred: it presupposes that all relevant DNA profiles have been completely detected. It is apparent that this can lead to problems. Suppose one of two alleles has dropped out at one of twelve loci, so that eleven loci result in a match between questioned and known profiles, but one locus does not match. The probability that this has occurred by chance is vanishingly small, but absent some way of assessing the likelihood that dropout has occurred, how can one appropriately take this into account? No analyst would ever classify this as an exclusion, but simply disregarding the aberrant locus is to ignore evidence pointing in the opposite direction. The result is to introduce an inevitable element of subjectivity into the process.

6.5.2 Semi-Continuous Models

In *semicontinuous models* alleles are either present or absent in the sense that we admit the possibility that a peak may either represent a true allele, or be a stutter artifact, or be spurious; and that a peak that ordinarily should be present may not be. Semi-continuous models however do not draw on peak heights in the modeling process. Some software products that are semicontinuous based include Lab Retriever, David Balding's LikeLTD, and LRmix Studio (used in Europe); these all work reasonably well, and are all open source.

Here are some brief details about how these work; the fact that they are open source means it is relatively easy to find details online.

- The electropherogram must first be qualitatively evaluated
- Peaks are identified by the analyst as alleles or stutter
- Peak height information is not used.
- Probabilities are assigned to the possibilities of dropout and drop-in.

The failure to use peak height information simplifies the models but at some cost.

6.5.3 Fully Continuous Systems

These are being rapidly adopted throughout the forensic community, thanks to available implementations. The two major software products used in the US are STRmix and TruAllele, in which peak height is a continuous variable. For concreteness, the description here will focus on STRmix.

The key idea is really very simple. Suppose one is analyzing the issue of peak height balance, and consider the question, do two peaks correspond to the two alleles coming from a heterozygous individual at a locus, or two alleles coming from two different individuals? Here is a simple toy model to think about this question. Let us model peak heights as normally distributed (that is, the famous bell-shaped curve), and assume that in the case of a heterozygote, the heights of its two peaks follow the same normal distribution $N(\mu, \sigma^2)$, with the mean μ and standard deviation σ being model parameters to be estimated from the totality of the data across all loci. If X and Y denoted the observed peak heights for the two alleles, then $X - Y$ is the observed difference in peak heights, having the distribution $N(0, 2\sigma^2)$ and estimating σ enables us to judge how likely or unlikely the observed difference $X - Y$ is. If, on the other hand, the two peaks come from two different individuals, then X will be distributed $N(\mu, \sigma^2)$, Y distributed $N(\nu, \tau^2)$, and $X - Y$ will have the distribution $N(\mu - \nu, \sigma^2 + \tau^2)$, and we can again judge how likely or unlikely the observed difference $X - Y$ is under these assumptions. The likelihood ratio will compare these two quantities, and provide a factor in favor of one of the two competing scenarios. The point is that in comparing the two explanations we must contrast how well they "fit together"; that is, although one hypothesis as to the source may fare poorly at one locus, it may do well at others, and this overall synergy may point strongly in one direction or another.

The ultimate challenge, as previously discussed, is that limited or degraded quantities of DNA and technical artifacts of the PCR copying processes complicate interpretation. Fully continuous systems deal with such problems by modeling peak heights, attempting to adjust for what is essentially noise in the system. Key model elements include:

- *Contributions* to a peak are assumed to be *additive*.
- *Variability* about *expected* peak heights
  - is *modeled* using empirical data
  - assume *variability* is large for small peaks and small for large peaks
- *Independence* of peaks at and between loci is assumed.

A key parameter is what is termed *mass*, namely the amount of DNA present for analysis, the result of template, degradation, amplification inhibition, and other factors. Mass is determined by the *mass parameters*

- t_n: *template*, amount of DNA from contributor n
- d_n: *degradation*, models decay of product
- A^l: *locus offset*, models amplification level at a locus

The mass present is a result of all of these effects. If there are N contributors, L loci, then there are $2N + L$ parameters to estimate. The result is a complex model having a large number of parameters that must be simultaneously estimated.

Let's look at these a bit more closely. First there is *template*, involving one empirical and one modeling assumption. The empirical assumption is that peak height is *approximately linear* in the template (e. g., twice as much template means a peak twice as high). The modeling assumption is that peak height (or area) is assumed additive (= "stacking" in the US). Estimating template based on these assumptions allows one to predict peak height; this is the basis of parameter estimation, matching the estimated peak heights with the ones actually observed.

Next there is *degradation*, basically anything that decreases the amount of template. (One complication here is that this can vary from one contributor to another.) Degradation manifests itself in a "ski slope": larger (longer) alleles are more prone to degradation (which is only to be expected, since the longer the repeat, the more DNA there is available to be degraded). Multiple causes for such degradation include chemical, physical, and biological "insults" (environmental factors) such as humidity, bacteria, and ultraviolet light.

Finally there is *locus specific amplification efficiency*, reflecting the fact that template and degradation do not completely account for variation. In STRmix amplification efficiency is modeled by a mean zero lognormal distribution, using a fixed but optimized variance determined by laboratory calibration data. (Other systems can and do use other statistical models.)

The result is a highly complex statistical model with so many parameters that the sophisticated methods of MCMC (Markov chain Monte Carlo) must be used to estimate them. (European and some US systems typically do not use MCMC.) There are, of course, disadvantages to the use of MCMC, which is not only compute intensive and makes considerable intellectual, mathematical, and technical demands on the user, but which also employ a form of stochastic optimization so that one does not get exactly the same answer each time the system is run on the same data (which can present challenges in explaining the results to the intended audience). The use of these systems is also subject to certain restrictions. In particular, they work much more slowly as the number of contributors increases and usually cannot handle more than four contributors.

6.6 Conclusion

The use of DNA identification evidence justly remains the gold standard for current forensic use. But its logic is complex and the entire field rapidly evolving. This short paper has attempted to give a sense of both the rewards and challenges resulting from adopting what is effectively a highly sophisticated form of statistical inference.

For further information, see the excellent introduction of Evett and Weir (1998), and for more comprehensive treatments, Butler (2015) and Buckleton et al. (2016).

References

Buckleton, J.S., J.A. Bright, and D. Taylor. 2016. *Forensic DNA Evidence Interpretation*. 2nd ed. Boca Raton: CRC Press.

Butler, J.M. 2015. *Advanced Topics in Forensic DNA: Interpretation*. San Diego: Academic Press.

Curran, J.M., et al. 1999. Interpreting DNA Mixtures in Structured Populations. *Journal of Forensic Sciences* 44: 987–95.

Evett, I.W., and B.S. Weir. 1998. *Interpreting DNA Evidence*. Sunderland: Sinauer.

Kaye, D.H. 2013. The Genealogy Detectives: A Constitutional Analysis of "Familial Searching". *American Criminal Law Review* 50: 109–153.

NRC1. 1992. *DNA Technology in Forensic Science*. Washington: National Academies Press.

NRC2. 1996. *The Evaluation of Forensic DNA Evidence*. Washington: National Academies Press.

Phillips, C. 2018. The Golden State Killer Investigation and the Nascent Field of Forensic Genealogy. *Forensic Science International Genetics* 36: 186–188.

Wambaugh, J. 1989. *The Blooding: The True Story of the Narborough Village Murders*. New York: William Morrow.

Weir, B.S., et al. 1997. Interpreting DNA Mixtures. *Journal of Forensic Sciences* 42: 213–222.

Zabell, S. 2005. Fingerprint Evidence. *Journal of Law and Policy* 13: 143–179.

Chapter 7
Reasoning With and About Bias

Chiara Manganini and Giuseppe Primiero

Abstract The widespread emergence of phenomena of bias is certainly among the most adverse impacts of new data-intensive sciences and technologies. The causes of such undesirable behaviours must be traced back to data themselves, as well as to certain design choices of machine learning algorithms. The task of modelling bias from a logical point of view requires to extend the vast family of defeasible logics and logics for uncertain reasoning with ones that capture some few, fundamental properties of biased predictions. However, a logically grounded approach to machine learning fairness is still at early stages in the literature. In this paper, we discuss current approaches to the topic, formulate general logical desiderata for logics to reason *with and about* bias, and provide a novel approach.

Keywords Machine learning · Data bias · Algorithmic unfairness · Non-monotonic logics · Uncertainty

7.1 Introduction

A broad concern, now ubiquitous in both the technical and the philosophical literature, regards the trustworthiness of machine learning systems. The question is even more challenging in view of the opaque nature of these models, which often prevents us from precisely knowing or examining their inner structure (Linardatos et al. 2020).

A possible strategy to assess trustworthiness in these opaque settings is to check the actual behaviour of the model against a desirable one. Logics have recently been designed to formalise the trustworthiness of probabilistic programs and to reason about them, with a specific focus on determining statistical distance measures for

C. Manganini (✉) · G. Primiero
Logic, Uncertainty, Computation and Information Lab & PhilTech Research Center for the Philosophy of Technology, Department of Philosophy, University of Milan, Milan, Italy
e-mail: chiara.manganini@unimi.it; giuseppe.primiero@unimi.it

H. Hosni, J. Landes (eds.), *Perspectives on Logics for Data-driven Reasoning*, Logic, Argumentation & Reasoning 35,
https://doi.org/10.1007/978-3-031-77892-6_7

such systems with respect to their desirable behaviour (D'Asaro and Primiero 2021; D'Asaro et al. 2023; Genco and Primiero 2023). A specific formulation has also been offered to model classifiers whose incorrect output could be due to forms of bias, using a non-symmetric distance that reflects the systematic skewness of the result (Primiero and D'Asaro 2022). In this sense, such logics offer verification and reasoning tools for the trustworthiness of black-box models with respect to explainable surrogate counterparts, such as those obtained by symbolic knowledge extraction (SKE) (Calegari et al. 2020; Calegari and Sabbatini 2022).

Within this broad debate, the specific problem of machine learning bias has attracted significant interest in recent years, and the issue has been raised vigorously in a number of studies. To name a few: Buolamwini and Gebru (2018) presents algorithmic discrimination based on gender and race; Mehrabi et al. (2021) lists different types of bias in AI applications; Miceli et al. (2022) investigates the power dynamics and the political nature of data in AI; Eubanks (2018) illustrates the emergence of bias in cases of AI introduction for welfare, housing, and child protection in the US. Several tools have been developed and deployed to mitigate bias, also for explainability purposes: AIF360,[1] LIME,[2] or the recent BRIOxAlkemy,[3] see Coraglia et al. (2023). But the design of symbolic approaches to resolving bias for AI systems, in particular through appropriate logical systems, is still in the early stages. In Sect. 7.2.2 we provide an overview of some initial works on the topic.

Two questions can be considered essential for the aim of developing logical systems to mitigate the bias problem in AI:

Problem 7.1 (Reasoning with Bias) How does one reason validly from possibly biased information?

Problem 7.2 (Reasoning About Bias) How does one reason validly about possibly biased systems, to determine conditions and limits of their results?

A preliminary formulation of our approach was offered in Manganini and Primiero (2023). In the present version, we significantly revise and extend our analysis to provide criteria for understanding what is required to appropriately formulate the two problems. In order to do so, in Sect. 7.2 we start with an overview of current works in sub-symbolic and symbolic approaches to bias identification in machine learning. In Sect. 7.3, we present some general desiderata for a logic designed to

---

[1] https://github.com/Trusted-AI/AIF360.

[2] https://github.com/marcotcr/lime.

[3] https://github.com/DLBD-Department/BRIO_x_Alkemy.

reason *with and about* bias. In Sect. 7.4, we propose a logical formalisation of the notion of biased machine learning system and show how it accommodates the mentioned desiderata. We then turn to the question of how to quantitatively assess the presence of bias in Sect. 7.5. Finally, we explain how this logical framework can be applied and expanded in the future in Sect. 7.6.

7.2 Related Works

7.2.1 Background: Bias and Fairness in Machine Learning

When it comes to *qualifying* machine learning bias, a number of nomenclatures can be found in the literature, most of which tend to categorise biased predictions in terms of qualitative differences in the causes engendering them. Among the statistical factors most discussed responsible for bias, we find nonrandom sampling of subgroups (hence producing a sampling bias), the lack of diversity of the sample (representation bias), and the distortions that can emerge from the aggregation of datapoints (aggregation bias) (Mehrabi et al. 2021; Suresh and Guttag 2019; Olteanu et al. 2019; Pombal et al. 2022).

However, this approach might be quite dispersive. As Mehrabi et al. (2021) and Suresh and Guttag (2019) themselves emphasise, bias can occur in any phase of the machine learning pipeline, from data collection and processing (data bias), to the training of the learning algorithm (algorithmic bias), its deployment into a real and open environment and its integration with human decision makers (user interaction bias). Even worse, in many real-world scenarios, the prediction process has feedback loops, which means that the model decisions determine which data can be observed after the predictions (selective labels problem De-Arteaga et al. 2018; Lakkaraju et al. 2017).

A more mathematically oriented and complementary approach to the problem of bias focuses on its *quantitative* assessment in actual predictions, by the means of an adequate set of measures. Central to this family of approaches is the notion of "protected attribute", defined as an inherent or acquired characteristic of the individual—such as their race, gender, age, economic status, etc.—which may be of particular importance in decision-making for social, ethical, or legal reasons.

Generally, the quantitative approach to bias aims to measure how much the protected attribute has illegitimately influenced the predictions of a system. We can broadly distinguish between observational and causality-based measures of fairness (Castelnovo et al. 2021). As opposed to the latter, observational measures quantify bias (or equally, fairness) only based on the observable distributions of the predictions, therefore requiring little extra knowledge on the causal structure underlying the data. Among observational measures of fairness, it is standard to distinguish between group (or statistical) and individual (or similarity-based) metrics (Castelnovo et al. 2021; Verma and Rubin 2018; Chouldechova and Roth

2018). On the one hand, group fairness evaluates bias in terms of disparity of various statistical measures across groups of individuals of the test sample, based on the protected attributes. Conversely, individual measures quantify bias as a disparity in the predictions of pairs of similar individuals (according to a chosen similarity criterion), following the intuition that fairness occurs when similar individuals are treated similarly.

An orthogonal distinction for observational measures of fairness is between those that require knowledge about the true labels (ground truth) of the inputs in order to be computed, and those that remain blind with respect to it (Verma and Rubin 2018). The former capture a notion of fairness of prediction that is grounded in the actual adequacy of the prediction itself, and are therefore called measures of "fairness in impact", while the latter focus on a notion of "fairness in treatment", intended as an undesired correlation between protected attribute and prediction returned by the model (Gajane 2017).

Our research below proceeds in a direction related to what recently proposed in work by Sharma and colleagues (Sharma et al. 2020, 2021). They propose a new group of fairness metrics for classification tasks based on the concept of "burden", which quantifies how difficult it can hypothetically be to obtain a favourable outcome for a specific group G of individuals. The motivating intuition is that, if a considerable disparity in the average burden across groups is found, the model is making significantly more difficult for some groups to obtain the favourable outcome. The burden of a single input is calculated counterfactually, exploiting standard techniques for counterfactual explanations: for each member of G who obtained an unfavourable outcome, a counterfactual input is found that would have obtained the favourable outcome instead. Then, a distance measure is computed between the input and its counterfactual according to a suitable similarity criterion.

As will become clear, the general idea behind the burden measure seems particularly in line with our proposal of a correction distance (introduced in Sect. 7.5) to quantify how far the system is from correcting a wrong classification. Based on this measure, a group-based and an individual-based fairness metric will be then defined. At a methodological level, however, our approach is quite different from that of Sharma et al. (2020): it is proof-theoretic rather than mathematical and makes use of feature importance rather than counterfactuals. Interestingly, though, a recent work Albini et al. (2023) establishes an equivalence between counterfactual explanations and game-theoretical feature attribution methods.

7.2.2 Logical Formalisations of Bias and Outline of a New Proposal

From a logical point of view, the task of finding adequate fairness measures can be seen as subsumed under the broader task of applying verification methods to machine learning systems. For example, Seshia et al. (2018) proposes a list of the robustness and safety properties that deep learning systems should satisfy. In this spirit, a few works have recently attempted to implement an actual formalisation of fairness and bias usable as specifications for formal verification. Let us briefly discuss them.

Kawamoto (2020) formalises the properties of supervised machine learning models through a statistical epistemic logic. Distributional Kripke models are used to define statistical measures over the model predictions. Fairness is measured by the means of a modal operator of "conditional indistinguishability", by the means of which well-known statistical measures of fairness are translated in terms of degree of indistinguishability between the sub-groups of the test sample.

Belle (2023) significantly expands the scope of this framework, by formalising fairness and bias in a dynamic epistemic setting suitable to logically capture bias in the presence of actions, beliefs, and plans.

In contrast, Ignatiev et al. (2020) focus on assessing the presence of data bias in the training sample used for logic-based algorithms (such as decision trees and decision sets). A test is presented to decide the fairness of the training dataset, based on an individual fairness measure inspired by the principle of Fairness Through Unawareness (Verma and Rubin 2018). As such, the work focuses on a notion of fairness which is neither in impact nor in treatment. Rather, fairness is here intended as a property of the training dataset. For this very reason, however, the metrics proposed can only partially capture the presence of machine learning bias, since the training data is not the only responsible for discriminatory predictions (see Sect. 7.3.3). Nonetheless, the paper provides a valuable discussion of the theoretical desiderata for a criterion of fairness. Interestingly, among a number of requirements (coding independence, lack of arbitrariness condition, simplicity, monotonicity), the authors notice that an appropriate fairness measure should be invariant under the addition of irrelevant protected features. Additionally, they recognize that their measure falls short of meeting this requirement. In this regard, it is worth remarking that although the approach we propose in the present work is different because we want to measure fairness in impact and not in training, it will be easy to see that our proposal can accommodate the intuition behind the principle of invariance under irrelevant features.

Finally, Liu and Lorini (2023) introduce a modal language to formalise binary classifiers and their properties, and for which they provide axiomatics and complexity results. With this, not only they formalise a variety of relevant notions of explanation (abductive, contrastive, counterfactual), but also propose a logical formalisation of the principle of Fairness Through Unawareness already mentioned. Overall, their proposal seems to be the most detailed existing account of how to reason *about* bias, according to the distinction between Problem 1 and Problem 2 we made above.

However, it should be noted that none of the mentioned approaches engages with the task of formalising the inferential engine that underlies a biased outcome, that is, the problem of how to reason *with* bias. The novelty of our contribution lies in the fact that we approach the phenomenon of bias at this level as well. More precisely, we first address Problem 7.1, by sketching a deduction system that models the inferences made by a (trained) classifier predicting new inputs. In this framework, it is possible to syntactically model a process that remains largely unexplored in practice, namely that of prediction update, as the amount of information available and relevant for the prediction. It is worth noticing that under this reading the logical system is intended to be essentially non-monotonic. Although not fully developed, and only introduced in the form of a proof-theoretic system, we believe that this approach represents the basis for an interesting stance on logics to reason with bias. After that, we will turn to Problem 7.2, defining two new fairness measures that precisely rely on the amount of information used to answer to Problem 7.1.

7.3 Logical Desiderata for Bias

As a preliminary work, we start from an informal description of four properties of the bias phenomenon we want to formalise.

Desideratum 1: Our Logic Should Be Able to Model Incorrect Conclusions that Are Validly Deduced by the Classifier The informal concept of bias we want our logic to capture is that of bias *qua* prediction error of a peculiar type.[4] Although not completely unproblematic (Castelnovo et al. 2021), this working definition of bias has many advantages. First, it enjoys a broad, mostly implicit acceptance, at least in the broad literature on statistical measures of fairness based on false positives and false negatives (Verma and Rubin 2018). Moreover, this definition is statistically approachable, normative, and still largely neutral under a moral point of view. To reconnect with the taxonomy briefly summarised in 7.2.1, it is sufficient to say that, with our logic, (1) we want to capture the presence of bias quantitatively; (2) the notion of fairness we want to capture is grounded in prediction incorrectness and therefore requires knowledge of the true labels of the inputs (fairness in impact). This statement will be expanded in further detail in Sect. 7.3.2.

Desideratum 2: Our Logic Should Be Parametric with Respect to Training Data, Test Data, and Model Understanding bias in terms of prediction error entails that its emergence largely depends on the properties of the inputs used to test it. In this sense, a consequence or inference relation that simulates biased reasoning shall refer not only to training data and the model as its variables, but also to test data. More details will be given in Sect. 7.3.3.

Desideratum 3: Our Logic Should Be Able to Model How Potentially Incorrect Predictions Are Validly Corrected as the Amount of Information Available Relevant to the Classification Increases In the process of assigning features to a datapoint, an unbiased classifier would satisfy monotonicity, where the probability of some output cannot decrease when the input increases. On the other hand,

[4] Primary goal of the discussion in Sect. 7.3.2 will consist in conceptual clarification of these very properties.

reasoning in the presence of bias requires some form of non-monotonic reasoning, where an initially incorrect prediction should be corrected provided the right amount of information. Different behaviours with respect to such correction strategy for distinct (groups of) individual(s) denote therefore a biased stance of the classifier. This requirement will be further justified and formalised in Sect. 7.3.4.

Desideratum 4: Our Logic Should Be Able to Accommodate a Minimal Qualitative Distinction Between Biases Bias comes in many forms. As already said, a qualitative categorisation of bias risks ending up incomplete. For this reason, we only require to capture a minimal distinction between types of biases in our logic: biases engendered by a too limited amount of information on the one hand, and biases caused by an incorrect relevance assignment to features on the other hand. More on this in Sect. 7.3.5.

In what follows, we will justify and formalise each of these four intuitions in more detail. Before doing so, let us begin with an explanatory toy example.

7.3.1 A Toy Example

Consider the following scenario.

Example: Perception of a Figure
Bob is walking on the street. On the other side, he sees a figure: slim, long blond hair, he thinks he is seeing his friend Alice. But not being sure, he starts walking towards her. At each step, he is closer to the subject, his perceptions are clearer and he is able to refine his confidence in the correctness of his judgement. Approximately halfway, Bob has acquired all the visual information sufficient to realise that he was wrong. The features of the figure misled him in thinking it was Alice, as the subject on the other side of the street is a young boy he does not know.

Few elements are worth considering in this short, simplistic story. First, Bob's wrong judgement is not casually occurring, but it is induced by some characteristics of the subject he is seeing. In particular, the same judgment would have occurred for any subject with the characteristics that Bob attributes to Alice (slim figure, long blond hair), and it denotes a prejudice towards a category of individuals (boys are not, at least not immediately, identified by those attributes).

Second, the wrong judgement emerges from Bob's previous experience: he probably never had a male friend with those characteristics, and among his female friends Alice must be the one most likely to be identified by them, possibly Bob's only female friend with long blond hair and a slim figure.

Third, Bob is making his judgement by collecting data as he proceeds toward the subject, but his "training" and prejudice fail him. At some point, having collected enough information, he realises that he was wrong, and corrects himself. In other words, there exists a specific step towards the figure at which Bob updates his judgment. This revision consists of a failure of monotonicity and it can either act on the premises ("blonde slim figures are mostly girls", "there is at most one slim blond girl I know") or on the rules of inference (Modus Ponens, or Weakening) of Bob's reasoning system.

Finally, note that a fully biased individual, or one unable to collect further information, would have stayed with the wrong judgement. Hence, the distinction between bias due to insufficient information (Bob stops halfway before identifying the subject as an unknown male) and bias due to wrong assignment of relevance to certain features (a blond-haired, slim figure is more likely to be a female, i.e., these features have higher probability of individuating a girl than a boy) are both crucial aspects of the problem.

7.3.2 *Skewness of Incorrect Predictions*

As already said, the working notion of bias we want to formalise in our logic is that of bias as a particular type of prediction error. In this section, we will spell out this requirement further, specifying which properties of prediction errors qualify them as occurrences of bias.

First, bias as we intend it consists of a systematic error of prediction, i.e., a consistent overestimation or underestimation of certain datapoints in the test sample. We further impose (1) that such a systematic error must be harmful to ("go against") a sub-domain of individuals, and (2) that group membership to this sub-domain is determined by a protected attribute (such as race, gender, age, or economic status).

Regarding condition (1), "against" is to be intended as "opposite to the direction given by the most favourable prediction". Such a direction of course varies in relation to the type of scenario considered. One broad, useful distinction is that between allocative and punitive contexts: in the former case, some beneficial resource (a job, a loan, a subsidy, etc.) is to be allocated to individuals based on a prediction; in the latter, some punitive action is to be imposed to individuals, based on some risk measures (fraud, recidivism, criminal risk, etc.).

In the case of allocative contexts, bias intended as systematic, harmful, and partial prediction error emerges as a disparity in false negatives across groups, harmful for the protected group. In other words, those individuals who belong to the protected group and should in fact profit from a resource are proportionally more likely to be predicted as not qualified to obtain it. Specularly, in the case of punitive contexts, bias emerges as a disparity in false positives across groups, harmful for the protected group. In other words, innocent individuals who belong to the protected group are more likely to be predicted as risky for committing a certain crime.

Example: Fraud Classifier
A classifier D is used to predict the risk of insurance fraud. To each new entry in a set of insurance claims, it either assigns the target label "Fraud", if the predicted risk is above a certain fixed threshold, or "¬Fraud" otherwise. D uses the protected binary feature "MalePolicyholder" for its prediction.

Suppose that a new claim a is to be classified. At an early stage, the only available information is $\neg MalePolicyholder(a)$, based on which D predicts $Fraud(a)$. However, when D is fed with a second piece of information about a, the classifier lowers its assessment value and predicts $\neg Fraud(a)$, which eventually turns out to be the correct label for a.

A formal representation of a reasoning system that accounts for bias will need to model this skewness—intended as systematicity, harmfulness, and partiality—of incorrect predictions within a logical inference relation $\mid\!\sim$. To come back to the example above, the reasoning system modelling Bob's prediction will assign to a blonde, slim individual a higher probability to be female.

7.3.3 Dependency on Data and Model

A second property of bias that we wish to capture is its dependency on both data and the model. In the literature it is well acknowledged that a skew in frequency of the classes in the training data leads to disparate error rates on the underrepresented attributes (Suresh and Guttag 2019). However, what is less recognised is that training data is not the only place where bias can lurk. As pointed out by Mehrabi et al. (2021) and Suresh and Guttag (2019), also algorithm design choices concerning loss function, optimiser, hyper-parameters, and even more subtle ones (like learning rate and length of training) can impact fairness due to the fact that underrepresented features are learnt later in the training process (Hooker 2021).

The actual occurrence of bias as defined in the previous Section, however, depends on the test sample. Sometimes biased models do not produce discriminatory results at all, due to contingent conditions. Think of a situation where an imbalance in the frequency of the gender classes in the training sample causes the model to have the disposition to produce biased outcomes against women. Nonetheless, it is possible that no biased prediction is actually made, as long as none of the inputs of the test sample triggers this disposition, i.e., as long as the model is not tested on women. In this sense, although the presence of bias presence depends on training data and model choices, its emergence depends on test data.

Example 7.1 (Example) Reconsider the gender-biased learning algorithm D for the detection of fraud risk. Imagine now that, due to poor selection of inputs for the testing process, the totality of the test inputs are male, i.e., each of them satisfies the predicate "Male". What can we say about a possible gender bias of this algorithm? On the one hand, D has the disposition to produce biased outcomes against women; on the other hand, no gender-biased prediction is actually produced, since this disposition is never exhibited.

Our formal modelling will have to be parametric with respect to these three variables: training data, model, and test data. More precisely:

- The training data will be modelled as a parameter of the consequence or inference relation of the system. It will refer to a set of predicates and a domain of constants from which one assumes to have made inferences before, thereby applying the correct inferential steps: $\mathcal{S}^{train} := \{\mathbb{P}^{train}, \mathbb{D}^{train}\}$.
- The model will be formalised as the relation $\mid\!\sim^{trained}$, which carries information about the relevant set of inputs used in the training process, as defined by $\mathcal{S}^{train}$. Note that in applications and especially for verification purposes, this information might be inaccessible or opaque ($\mid\!\sim^{\emptyset}$).
- The test data will be expressed as a distinct set of predicates $\mathbb{P}^{test}$ and constants $\mathbb{D}^{test}$, to which the inference relation must apply. Errors such as out-of-distribution and bias in historical data will emerge as assumptions about their comparison with $\mid\!\sim^{trained}$.

7.3.4 Non-Monotonicity

In Bob's story, there is a sense in which getting closer to Alice's lookalike allows Bob to make his perceptions more and more precise. At each stage, Bob's might assign a degree of certainty on the person's identity. An increase in accuracy or in available information can lead to a change of classification, up to the point where a degree of certainty is sufficient to assert $P(a)$ or $\neg P(a)$. At the beginning of this process, the total-distance situation from Alice's lookalike may be idealised as a "minimal-knowledge" condition, in which Bob has exactly one piece of information to make his prediction, with the minimum accuracy possible, and hence a minimal degree of certainty.

Out of the metaphor, our example translates as follows.

Example: Threshold for Fraud
The system D is fed step by step with features from a predetermined list that contains age, gender, address, etc. about datapoint a. At each step, D must classify a. Imagine now that at step m of this process, D assigns an output

(continued)

value that, according to the rules of the algorithm, implies the target feature "Fraud" in the case. However, the information provided in step $m + i$ makes D revise the confidence in this prediction under a certain threshold, and the system outputs value "$\neg$ Fraud".

We will model initial data on the test sample (how much Bob is able to evaluate from the other side of the street) with an assignment of how much the current information is (estimated to be) with respect to the total amount of information required to make a correct evaluation. For example, Bob might estimate that he would need only to get halfway from the person in question to be sure about her identity, and estimate this location to be 10 steps away, while being currently at $1/10$ of such distance. Or Bob might think he only needs to assert correctly some features of the person in question, while having only some such information, thereby granting himself some degree of confidence.

On the other hand, the zero-distance situation from Alice's lookalike is a sort of "omniscience" condition, in which Bob holds all the available pieces of information, with the maximum accuracy possible. We can assume that the latter situation is always associated with Bob's correct classification (ground truth). This will require decorating formulas with values for information available and the associated confidence rate.

To model the behaviour of a learning system that corrects its own prediction on a new input a as the amount of available information increases we need to establish inferential rules to approximate a perfect classifier even in conditions where the classifier on the test data produces contradictory result with respect to the ground truth. Hence, non-monotonicity of the inferential relation is essential for clarifying the relationship between a biased outcome and the information used to generate it, both qualitatively and quantitatively.

This intuition is expressed in the following principles:

Proposition 7.1 *A well-trained classifier should produce correct classifications (within admissible error margins) when provided with all available, correct information* $N = 1$.

Proposition 7.2 *A well-trained classifier should produce correct classifications (within admissible error margins) when provided with sufficient, correct information* $N \leq 1$.

Proposition 7.3 *Any classifier with information* $M < N$ *may provide an incorrect classification.*

Note that, in practice, a classifier assumes $N = 1$ to be the normalised value of the whole set of features it uses for the prediction, although it could be taken to be a smaller value (i.e. the prediction can be correct even in absence of some information, or when the information available is of lower weight than the total). And $M < N$ is always the weighted value of the set of features used for an incorrect prediction. When incomplete information is already sufficient for a correct prediction, then $M < N \leq 1$.

With this in mind, let us now consider a machine learning system, which we know to be biased, according to the analysis in Sects. 7.3.2 and 7.3.3. In other words, this means that:

- the system generates systematic (i.e., with a net direction) erroneous predictions against a certain protected attribute (i.e., individuals with the protected attribute), and
- the errors are caused by the data on which it has been trained or by the design of the model itself.

In this scenario, we are interested in exploring how the presence of bias in the system quantitatively depends on the amount of information used to return an incorrect prediction. A first intuition is that the system may be considered *maximally* biased against a certain protected attribute when, given a new datapoint with that attribute and for which a wrong classification has been predicted, no additional information allows to correct it, i.e. for any amount $M \lesseqgtr N$. Conversely, the system may be considered *minimally* biased when, given the same datapoint, when a single additional piece of information allows to correct it. Note that, in this light, in order to evaluate bias against a protected class, it is sufficient to show that M is significantly higher for the members of the protected group, with respect to the non-members. We will offer such a comparative measure in Sect. 7.5.

To recapitulate, in the presence of a biased system, we expect our logic to:

- evaluate classifications with a degree of certainty, based on the amount of information available;
- measure the distance of any such incorrect classification from the correct one, i.e., inferred from information sufficient for a correct classification;
- assess how much information is required to fail monotonicity, i.e. to revise a classification assumed to have an incorrect result, and use the latter as a proxy of the amount of bias affecting the system when such measure is different for a protected class compared to any other class.

7.3.5 A Minimal Distinction Between Types of Bias

A minimal distinction between types of bias can be traced on the basis of this quantitative approach:

1. biased predictions produced by an insufficient amount of information available to the system to generate the prediction (sometimes called *selection bias* or *omitted variable bias* Mehrabi et al. 2021; De-Arteaga et al. 2018). Should Bob only check for hair length and colour, and shape of the body? Could the type of clothes or the company of the person in question be used as predictors as well?

2. biases produced by an incorrect assignment of relevance to the features used for the prediction.

Our formal modelling will assess such criteria in terms of the training sample $\mathcal{S}^{train}$ and the test sample $\mathcal{S}^{test}$, how their predicate sets differ and how such comparison affects the graded evaluation made under the inference relation $|\!\sim^{trained}$ representing the trained classifier.

An important remark concerns the amount of information we assume to be available for modelling the classifier through our inference relation. On the one hand, it seems reasonable to consider the modelling of the classifier behaviour as obtained, for example, by an interpretability technique providing features' weights. In this case, the object of analysis by our inference relation would be the interpreted classifier, with all the available information. In this setting, reasoning *about* its behaviour seems to allow us to use all the information about the data set and the relevance of its features.

On the other hand, when using our inference system to simulate the behaviour of the classifier itself, we are potentially working in conditions of opacity, as information on the weights of the features is not necessarily available. Even under these stricter conditions, reasoning *as* the classifier is still possible. The inference system may be in fact entirely opaque with respect to the weights of the features, as long as it is transparent concerning the certainty value for the result of the classification. The inferential process and the mechanism that we will define to assess bias act only on these certainty values, allowing to assume nothing explicitly on how they are obtained by the associated weights, and by which function.

Notice, moreover, that the same considerations apply to the case in which the extraction of feature importance by some interpretability techniques is problematic for some reason. As an example, some issues notoriously arise in attributing SHAP values to one-hot encoded categorical features, since it is difficult to understand how the original categorical feature contributes to the overall prediction. More in general, some works Marques-Silva (2023) and Huang and Marques-Silva (2023) have recently challenged the adequacy of Shapley values as the game-theoretical underpinning of feature attribution methods. It is in fact shown that, for some predictions, significant discrepancies exist between the Shapley values of the features and their associated relevance, independently obtained through a feature selection method. The results seem to suggest that Shapley values are ultimately unable to capture the true relative importance of features.

In all these cases, the assumption of a unique consistent set of weights for the resulting classification may look too strong. Accordingly, we may bypass the feature importance attribution process and still work on the confidence rate of the classification.

7.4 A Logic to Reason with Bias

In this Section, we sketch the basic elements to define a proof-theoretic approach to logics to reason with biased premises, according to the elements illustrated in the previous sections. For reasons of clarity, our language will be used to model the standard case of supervised classification with a single, binary protected attribute and (possibly) many binary non-protected attributes. We leave the development of an appropriate semantics to another occasion. In the proof theory, we interpret the non-monotonicity desideratum discussed in Sect. 7.3.4 as a graded (i.e. probabilistic) derivability relation implementing the failure of the structural rule of Weakening.

We start introducing the following sets of predicates (also called attributes or features), upon which we will then define the Training and the Test Samples:

- $\mathbb{PR} = \{Q\}$ as the singleton containing the binary protected predicate.
- $\mathbb{NPR} = \{R_1, R_2, ..., R_n\}$ as the set of non-protected predicates.
- $\mathbb{T} = \{P\}$ as the singleton containing the binary target predicate.

Example: Loan Eligibility
A classifier is used to predict applicants' eligibility for a loan. Therefore, $\mathbb{T} = \{Loan\}$, where $Loan(a)$ can be true (if a is eligible for a loan) or false (if a is not eligible). The system classifies on the basis of the following sets of predicates: $\mathbb{PR} = \{Male\}$, and $\mathbb{NPR} = \{FullTimeEmployed, HouseholdOwner, UniversityDegree, Married, AnnualIncomeOver50k, Over35yo\}$.

Definition 7.1 (Training Sample) A training sample $\mathcal{S}^{train} = \{\mathbb{D}^{train}, \mathbb{P}^{train}\}$ is such that:

- $\mathbb{D}^{train} = \{b_1, b_2, \ldots, b_n\}$ is a finite set of elements of a domain denoting datapoints of the training sample;
- $\mathbb{P}^{train} = \mathbb{PR} \cup \mathbb{NPR} \cup \mathbb{T}$

Definition 7.2 (Test Sample) A test sample $\mathcal{S}^{test} = \{\mathbb{D}^{test}, \mathbb{P}^{test}, w, <^{w}\}$ is such that:

- $\mathbb{D}^{test} = \{a_1, a_2, \ldots, a_m\}$ is a finite set of elements of a domain denoting the datapoints of the test sample;
- $\mathbb{P}^{test} = \mathbb{PR} \cup \mathbb{NPR}$
- $w : \mathbb{P}^{test} \mapsto [0, 1]$ is a total weighting function that assigns to each predicate a measure of relevance for the prediction. Here, we assume that weights are already assigned to features, abstracting away from the assignment process itself, be it obtained by extra knowledge or by standard techniques, such as Gini distance or permutation methods;

- $<^w \subseteq \mathbb{P}^{test} \times \mathbb{P}^{test}$ is an ordering on the predicates based on w so that $\langle P_i, P_j \rangle \in <^w$ iff $w(P_i) \leq w(P_j)$ for any $P_i, P_j \in \mathbb{P}^{test}$. Moreover, we take the weights to be normalised, i.e., that $\sum_n w(P_n) \leq 1$;

The language is then defined by predicative expressions closed under negation, conjunction, disjunction and implication.

Definition 7.3 (Language)

$$\begin{aligned}
&\phi := P(a)^w \mid \neg\phi \\
&\Phi := \phi \wedge \phi \mid \phi \vee \phi \mid \phi \rightarrow \phi \mid \\
&\psi := T(a) \text{ such that } T \in \mathbb{T} \text{ and } a \in \mathbb{D}^{test} \\
&\xi := r \in \mathbb{R}
\end{aligned}$$

A few clarifications are important about the definition of $\phi := P(a)^w$. First of all, it is assumed that $\mathbb{S}^{test}$ is a noiseless sample, i.e., that all its inputs are labelled in a way that fully reflects reality.[5] This means that $P(a)$ with $P \in \mathbb{P}^{test}$ is to read as "It is the case that individual a has property P" or "Individual a is P". Conversely, $\neg P(a)$ is to read as "It is not the case that individual a has property P" or "Individual a is not P". Note that we do not assume a generalised version of the Excluded Middle, as it is sensible to consider only properties which can actually be derived (or their negation can) as target feature from the current list of available predicates for the datapoint under evaluation. Moreover, the weighted inference relation we introduce requires a weaker than classical definition of negation. Both these aspects are expressed by the rules regulating the negation connective, namely *failure* and *safety* in Fig. 7.2. Furthermore, with respect to the value of w in the definition $\phi := P(a)^w$, we have that:

- a satisfies predicate P and $w = w(P)$ for the target ψ;
- a does not satisfy predicate P, and $w = 1 - w(P)$.

This can be clarified as follows: for any datapoint, we check the current list of predicates (features); knowing that a predicate holds for a datapoint will weight in the classification at hand for the value of the predicate; knowing that the predicate does not hold, we take the inverse of the weight of that predicate, which also allows for non-binary classes. Furthermore, for the sake of brevity, below we will use the following shortcut to refer to the weight associated to a given variable: $w_i := w(P_i)$ where $P_i \in \mathbb{P}^{test}$.

Finally, with the metavariable ψ we denote the result of the classification $T(a)$ for a predicate T that ranges over the class $\mathbb{T}$ of target predicates and with ξ we refer to a real value associated to the degree to which the target predicate can be inferred from the current feature set.

---

[5] Note that this assumption is not imposed to the training sample $\mathbb{S}^{train}$, which means that the algorithm could be trained on a noisy set, sometimes referred to as the "noisy labels problem" Fogliato et al. (2020).

Example 7.2 (Example) Suppose that it is known that being overweight is 60% relevant for predicting the onset of a cardiovascular disease in patients and that a specific preventive health care is guaranteed to those who have a probability higher than 50% of developing the disease.

In our logic, this corresponds to the following. Given that $w(Overweight) = 0.6$, overweight individual a, about whom no other information is available, is predicted to satisfy also $Disease(a)$ with 60% of probability and therefore receives the preventive treatment. Conversely, under the same knowledge conditions, non-overweight patient b is predicted to satisfy $Disease(b)$ with a significantly lower (though not null) probability of $40\% = 100 - 60\%$.

7.4.1 A Proof-Theoretic Interpretation

We now present a graded proof-theory to model the classification of a test sample. In order to do so, we first introduce two simple rules to inductively construct a feature list, thereby defining a method that selects one feature at a time, coupled with its weight (See Fig. 7.1).

The induction starts from a list containing a single element, namely the protected predicate for the given datapoint (base case). At any step, an atomic predicative formula of weight w is added (selection), where the order in which predicates are added may depend from the availability of incoming information, or could be established by design. The overall weight of a feature list is denoted by $W = \sum(w_1, \ldots, w_n)$, as it is the sum of the weights of the predicates in it. Each selection allows to infer a new feature list, where a new weight w is added to the previous sum of weights W, always checking the stopping condition that the summation on the overall weight $W \leq 1$, which is the total amount of available information on a datapoint. Note that this simplistic additive process on a predefined list of features only returns the overall weight of the selected features.

Let us now clarify how, at each step of the induction, the current list of selected features corresponds to the set of premises used by the inferential system to output a probabilistic prediction.

Definition 7.4 (Classification) A judgement:

$$\Gamma^{test,W} \mid\!\sim^{trained} \psi, \xi$$

$$\frac{}{\{Q(a) \mid Q \in \mathbb{PR}\} :: features}\ base\ case$$

$$\frac{\Gamma^{w} :: features \qquad R(a)^{w'} \notin \Gamma \mid R \in \mathbb{NPR} \qquad w + w' \leq 1}{\{\Gamma, R(a)\}^{W=w'+w} :: features}\ selection$$

Fig. 7.1 Construction rules for the inductive selection of features

$$\frac{}{\Gamma^{W}, \phi \mid\sim \psi, \xi} \; classification$$

where $\xi = f(W + w)$, for some suitable f

$$\frac{\Gamma^{W} \mid\sim \psi, \xi}{\Gamma^{W} \mid\sim \neg\psi, (1-\xi)} \; failure \qquad \frac{\Gamma^{W} \mid\sim \neg\psi, \xi}{\Gamma^{W} \mid\sim \psi \rightarrow \bot, \xi} \; safety$$

$$\frac{\Gamma^{W} \mid\sim \psi_1, \xi \qquad \Delta^{W'} \mid\sim \psi_2, \chi}{\Gamma^{W}, \Delta^{W'} \mid\sim \psi_1 \wedge \psi_2, \xi \cdot \chi} \; \mathrm{I}\wedge$$

where Γ, Δ are independent

$$\frac{\Gamma^{W} \mid\sim \psi_1 \wedge \psi_2, \xi \qquad \Gamma^{W} \mid\sim \psi_1, \chi}{\Gamma^{W} \mid\sim \psi_2, \xi / \chi} \; \mathrm{E}\wedge_R \qquad \frac{\Gamma^{W} \mid\sim \psi_1 \wedge \psi_2, \xi \qquad \Gamma^{W} \mid\sim \psi_2, \chi}{\Gamma^{W} \mid\sim \psi_1, \xi / \chi} \; \mathrm{E}\wedge_L$$

$$\frac{\Gamma^{W} \mid\sim \psi_1, \xi \qquad \Gamma^{W} \mid\sim \psi_2, \chi}{\Gamma^{W} \mid\sim \psi_1 \vee \psi_2, \xi + \chi} \; \mathrm{I}\vee$$

$$\frac{\Gamma^{W} \mid\sim \psi_1 \vee \psi_2, \xi \qquad \Gamma^{W} \mid\sim \psi_1, \chi}{\Gamma^{W} \mid\sim \psi_2, \xi - \chi} \; \mathrm{E}\vee_R \qquad \frac{\Gamma^{W} \mid\sim \psi_1 \vee \psi_2, \xi \qquad \Gamma^{W} \mid\sim \psi_2, \chi}{\Gamma^{W} \mid\sim \psi_1, \xi - \chi} \; \mathrm{E}\vee_L$$

$$\frac{\Gamma^{W}, \phi \mid\sim \psi, \xi}{\Gamma^{W} \mid\sim \phi \rightarrow \psi, \xi} \; \mathrm{I}\rightarrow \qquad \frac{\Gamma^{W} \mid\sim \phi \rightarrow \psi, \xi \qquad \Delta^{W'} \mid\sim \phi, \chi}{\Gamma^{W}, \Delta^{W'} \mid\sim \psi, \chi \cdot \xi} \; \mathrm{E}\rightarrow$$

Fig. 7.2 Rules for classification

denotes a classification where:

- $\Gamma^{test,W} := \{Q_1(a), \ldots, Q_m(a)\}$ is a list of atoms $Q_i \in \mathbb{P}^{test}$ over an $a \in \mathbb{D}^{test}$ as per Definition 7.2 built according to the rules of Fig. 7.1, and with overall weight W.
- $\mid\sim^{trained}$ is a non-monotonic inference relation which refers to a (known or unknown) training sample as per Definition 7.1, and whose semantics is defined by inference rules presented below in Fig. 7.2;
- ψ is a predicative atomic formula $P(a)$, where $P \in \mathbb{T}$ is the target feature, and $a \in \mathbb{D}^{test}$. It is therefore the actual prediction returned by the model for input a.
- $\xi \in [0, 1]$ denotes the confidence value that $\mid\sim^{trained}$ assigns to ψ being validly inferred from $\Gamma^{test,W}$, as determined by the inference rules. In other words, it denotes the probability that prediction ψ holds. Prediction ψ, ξ is returned as a binary classification if we fix a threshold value τ, such that the final output is $P(a)$ if $\psi, \xi \geq \tau$ and is $\neg P(a)$, otherwise. Finally, the probability ξ is correlated with the amount of information embedded in the set of premises $\Gamma^{test,W}$, i.e, $\xi = f(W)$. For simplicity, we can assume ξ to be obtained by the identity map on W, discounting for example any accuracy rate on $\mid\sim^{trained}$.

Example: Classification with Confidence
The judgement:

$$\{Blonde^{0.1}(a), Slim^{0.1}(a)\}^{test,0.2} \mid\!\sim^{trained} (Alice(a), 0.2)$$

expresses that, according to the classifier $\mid\!\sim^{trained}$, the features $Blonde, Slim$ assigned to datapoint a in the test sample suffice for a to be classified as Alice with a 20% confidence rate. Whether such degree is sufficient to ultimately justify this prediction depends on the choice of a cut-off point τ.

The rules in Fig. 7.2 define the derivability relation for our classifier. The $classification$ rule says that the target feature is derivable with degree ξ from a feature list of weight W. As already said, if we use the identity map between w and ξ, the two values will coincide, discounting any accuracy measure.

The two rules for negation are useful to model the distinction between punitive and allocative contexts introduced in Sect. 7.3.2. $failure$ establishes that, if a list of features derives a property for a datapoint with degree ξ, then the negation of that property for that individual will be derivable from the same list of features with degree $1 - \xi$. This should be the standard behaviour of classifiers in punitive contexts, when ξ is low: if the system predicts a property with a certain degree (say: a predictive justice system associates "being guilty" with a low score, e.g. 20%), it should infer its negation (e.g. "not being guilty" with a high score, namely 80%). On the other hand, $safety$ establishes that, if a feature is inferred with low probability (hence its negation can be inferred with a high probability), then the system would be contradictory in making that feature valid with high probability. This should be the standard behaviour of classifiers in allocative contexts when ξ is high: if the system does not predict the intended property with high degree (say: a welfare system associates "not being eligible for credit" with high score, e.g. 80%), then it should infer its negation (e.g. "eligible for credit" with a low score, namely 20%).

The rules for $\wedge$ and $\vee$ introduce and eliminate the two connectives by making standard arithmetic operations on the associated probabilities. The $I \rightarrow$ rule implements the Deduction Theorem: inferring the property ψ with score ξ, from a list of features containing ϕ with weight w, should make it possible to infer ψ with score ξ from Γ alone, once ϕ is assumed with the associated weight. Implication elimination, as usual, implements Modus Ponens: from the validity of an implication with score ξ and the assumption of the antecedent with score χ, one can infer the consequent with score $\xi \cdot \chi$. Note that, interestingly, only in this rule ϕ appears on the right-hand side of the inference, as inferred from the set of premises $\Delta^{W'}$. In other words, ϕ is predicted on the basis of a (possibly different) set of features $\Delta^{W'}$, and is carried over in the prediction of ψ.

To evaluate bias within such reasoning engine, we want to quantify the amount of information that would be needed for the system to correct a previously erroneous classification. To do so, we construct an external classifier in which the ground truth is inferred:

Definition 7.5 (Ground Classification) A judgement:

$$\Gamma^{test,1} \vdash^{ground} \psi, 1$$

denotes a classification where:

- $\Gamma^{test,1}$ denotes that premises Γ^{test}, as per Definition 7.4, embed the entire available information about an input, which amounts to 1 as said in Definition 7.2;
- $\vdash^{ground}$ is a monotonic inference relation that infers the ground truth;
- ψ, 1, like in Definition 7.4, is the prediction, this time inferred with full certainty.

Of course the one just defined is quite a demanding notion of ground truth, as it requires both the maximum quantity of information possible, and full certainty of the conclusion. It is nonetheless possible to liberalise this definition along both these dimensions. In fact, what we are interested to model through this judgment is not necessarily a perfect classifier, but rather an inferential system that, unlike system $\mid\!\sim^{test}$, confidently outputs the correct prediction with the minimum quantity of information possible. Hence:

Definition 7.6 (Weak Ground Classification)

$$\Gamma^{test,W} \vdash^{ground} \psi, \xi$$

The condition of initial incorrectness of the classifier's prediction under which we are reasoning, therefore, can be expressed by means of the following pair of judgements:

Definition 7.7 (Initial Condition)

$$\Gamma^{test,W} \mid\!\sim^{trained} \psi, \xi \qquad \Delta^{test,W'} \vdash^{ground} \neg\psi, \xi'$$

The intended correction can be expressed as failure of the structural rule of Weakening for the system. In other words, when the above condition of initial discrepancy between a given classification of input a and its ground truth obtains, correction is possible if there exists an amount of information W about a with the following property. Given a list of features Γ^W that let us infer ψ with probability ξ, there is an extension of Γ such that $\neg\psi$ becomes valid with an updated degree $\mid \xi' - \xi \mid$, see Fig. 7.3.

Such an extension might correspond to the evaluation of the entire set of features, thus reducing to Δ^{test}. Note that when $\xi = id(W)$, i.e. the score on the judgement derived is the result of the identity map on the weight W of the information used, then the absolute difference in score assigned to the judgement in the conclusion

$$\frac{\Gamma^{test,W} \mid\!\sim^{trained} \psi, \xi \qquad \Delta^{test,W'} \vdash^{ground} \neg\psi, \xi'}{\Gamma^{W}, \Delta^{W'} \mid\!\sim^{trained} \neg\psi, \mid \xi - \xi' \mid} \; correction$$

Fig. 7.3 Structural rule for correction

corresponds precisely to the additional information in terms of weight required to reach the correct conclusion starting from the wrong classification. In the following we will keep this assumption. Hence, when the *correction* rule is applied with a sufficient degree for ξ' (i.e. because of a feature in Δ with a sufficiently high weight), or a sufficient number of times, the degree of confidence $\mid \xi' - \xi \mid$ will increase above the threshold τ, and eventually the conclusion ψ will no longer be inferrable. This means that some premises on a given derivation tree of the form $\Gamma^W \vdash \psi, \xi$ were biased, in the sense that the weights of currently available data was either insufficient for a correct classification, or skewed in the sense that the available features weighted strongly in favour of a given prediction. In these cases, one has to add *selection* rules until all sufficient information is available for a correct classification, and thus eventually *correction* is satisfied.

Example: Derivation Tree
Consider the following derivation tree:

$$\frac{\{Y^{0.1}(a), W^{0.4}(a)\}^{test,0.5} \mid\!\sim^{trained} (Fraud(a), 0.5) \qquad \{\neg E^{0.7}(a)\}^{test,0.7} \vdash^{ground} (\neg Fraud(a), 0.7)}{\{Y^{0.1}(a), W^{0.4}(a), \neg E^{0.7}(a)\}^{test,1} \mid\!\sim^{trained} (\neg Fraud(a), 0.2)}$$

in which assuming it can be inferred with score 0.5 that the case is a fraud using the information that the driver is young and a woman, it infers that it is not a fraud with score 0.2 when the information is added that the person is not an expert driver (weighting as per the ground truth). Consider now the derivation tree:

$$\frac{\{Y^{0.1}(a), M^{0.6}(a)\}^{test,0.7} \mid\!\sim^{trained} (Fraud(a), 0.7) \qquad \{\neg E^{0.7}(a)\}^{test,0.7} \vdash^{ground} (\neg Fraud(a), 0.7)}{\{Y^{0.1}(a), M^{0.6}(a), \neg E^{0.7}(a)\}^{test,1} \mid\!\sim^{trained} (\neg Fraud(a), 0.0)}$$

in which assuming it can be inferred with score 0.7 that the case is a fraud using the information that the driver is young and male it infers with score 0.0 that it is not a fraud when the information is added that the person is not an expert driver weighting as per the ground truth).

One way to make explicit the difference between the conclusions of these two derivation trees is to consider the distance from correction for each. In the first case, taking 1 as the full information, and subtracting the score of the conclusion, we obtain that the distance to $\neg Fraud(a)$ is 0.8 in the former tree, while in the second is 1. This shows that it is harder to correct the judgement obtained in the first classification, than it is in the second.

This offers a first, tentative answer to what above we called Problem 7.1, regarding how to reason *with* bias. On this basis, we now want to offer a generalised version of the metric underlying the correction rule, for evaluating (1) how much additional information is needed to correct a prediction, and in particular (2) how much such measure is imbalanced across different groups of the test sample. With this step we move to Problem 7.2, or the problem of reasoning *about* bias.

7.5 Metrics for Reasoning About Bias

Given the judgement:

$$\{Q_1(a), \ldots, Q_m(a)\}^{f(w_1,\ldots,w_m)=M} \mid\!\sim^{trained} \psi, \xi$$

and fixed an aggregation function f over the weights of the predicates $Q_1, \ldots, Q_m$, M denotes the amount of information currently available and used for a wrong prediction ψ with a score ξ. Similarly, in judgement:

$$\{Q_p(a), \ldots, Q_n(a)\}^{f(w_p,\ldots,w_n)=N} \vdash^{ground} \neg\psi, \xi'$$

N expresses the amount of information sufficient to compute by the same function f over the weights of the predicates $Q_p, \ldots, Q_n$ the correct classification $\neg\psi$ with score 1 respecting the ground truth (or with $\xi' \geq \xi$). It is reasonable to assume that the score ξ' assigned by a classifier trained on the ground truth may converge to 1. Note that it is possible that $\{Q_1(a), \ldots, Q_m(a)\} \cap \{Q_p(a), \ldots, Q_n(a)\} \neq \emptyset$. In the following, we abbreviate with $S := (\Gamma^W, \mid\!\sim^{trained})$ the pair of weighted features set and trained model. Similarly, we abbreviate with $G := (\Delta^{W'}, \vdash^{ground})$ the pair of weighted features set Δ and classifier function which returns a classification that matches the ground truth.

Maintaining the assumption that the scores $\xi = f(W) = M$ and $\xi' = f(W') = N$, the difference $(N - M)$ expresses the amount of additional information that would be sufficient to correct the initial incorrect prediction, by applying the *correction* rule. This value can be further normalised by the accuracy of the system. Put differently, $(N - M)$ expresses the amount of additional information that would be sufficient for the Weakening rule to fail. This difference, therefore, is a measure of non-monotonicity, weighted on the certainty of the prediction. On this basis, we define a distance measure as follows:

Definition 7.8 (Correction Distance) Given $S := (\Gamma^W, \mid\!\sim^{trained})$ that infers a formula $\psi := T(a)$ such that $T \in \mathbb{T}$, $a \in \mathbb{D}^{test}$, the correction distance with respect to $G := (\Delta^{W'}, \vdash^{ground})$ that infers formula $\neg\psi$ is defined as:

$$C(S, \psi) = 1 - (N - M) * accuracy$$

i.e., the inverse of the amount of additional information sufficient for $\mid\!\sim^{trained}$ to correct its prediction on a, weighted on certainty (as expressed by the accuracy of the system, known from previous analyses).

Given our definitions of N and M—as the amounts of information sufficient to give, respectively, a correct and an incorrect prediction—the measure C formally expresses an evaluation of how far the system is from correcting a wrong classification.

Moreover, we further impose that $(N - M) > 0$ if we want to model the non-monotonic setting exemplified in Sect. 7.3.4, in which an increase of the available information is always assumed to result in an improvement (or, in a binary setting like ours, a correction) of the initial prediction (see Proposition 7.1). For this reason, the score on the correction rule is defined by an absolute difference. Moreover, when defining the order relation $<^w$ in Definition 7.2, we said that all the feature weights together sum up to 1. Therefore, $N \leq \sum_{P_i \in \mathbb{P}}^{n} W(P_n) = 1$.

Note, finally, that C can be interpreted as a probabilistic measure. For example, if $C(S, \psi) = 10\%$, we could say that the system will eventually reach the correct prediction with 90% probability.

Example 7.3 (Example) Recall the anti-fraud system mentioned in previous examples. Figure 7.4 shows how $C(S, \psi)$ varies in relation to M, N, and accuracy, assuming that $\Gamma^W \mid\!\sim \psi$ is an incorrect classification. On the x-axis we measure the available information M and on the y-axis the corresponding correction distance.

If we focus on the system denoted by the dashed yellow line, where the amount of information required for a correct classification is 60% of the total and accuracy is known to be 90%, in the virtual case in which no information is used for the

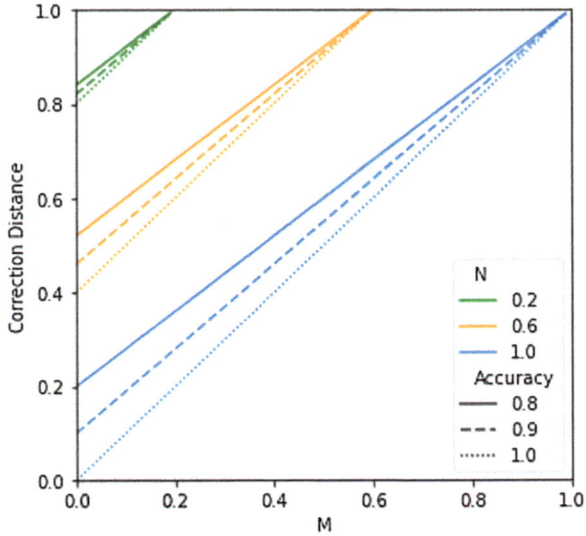

Fig. 7.4 Examples of correction distance for incorrect predictions, at different levels of N and Accuracy

prediction, the correction distance attributed to the classification is already set at 46%. The more the system proceeds with the acquisition of new information, say 40%, the correction distance attributed to a wrong classification rises to 82%.

For comparison, if we take a system with maximal accuracy (100%) and little information required to infer a correct classification (20%), a wrong classification without any information is already assumed to be 80% distant from its correction (dotted green line).

Finally, a system with low accuracy (80%) and which requires total information to infer correct classifications (100%)—like the one represented by the solid blue line—will be evaluated to be 36% distant from correction for the case of $M = 20\%$. While with $M = 90\%$, if a wrong classification is still being returned, its correction distance will be evaluated at 92%.

In the light of the above, intuitively a system is maximally distant from the correction of a wrong prediction $P(a)$ obtained based on amount of information M, if there is no additional amount of information M' such that the sum of $M + M'$ results in an amount of information sufficient for predicting correctly $\neg P(a)$. To formalise this, we use the correction distance to evaluate the amount of information required for the rule *correction* from Fig. 7.3 to hold.

Definition 7.9 (Maximally Distant-from-Correction System) Given $\Gamma^W \mid\sim^{trained} \psi, \xi$ and $f(W) = M$, then $C(S, \psi) = 1$ if and only if $\not\exists M'$ such that $f(W') = M + M' = N$ and $\Delta^{test, W'} \vdash^{ground} \neg\psi, \xi'$.

A maximally distant from correction system is therefore one such that no amount of correct information whatsoever can make it correct. Intuitively, when a system will behave as to be maximally distant from correctness with respect to any given property or (group of) individual(s), it will be deemed just to be an inaccurate classifier. On the other hand, a classifier which cannot be corrected provided novel data when acting on a given property or (group of) individual(s), *but not with respect to another*, can intuitively be said to be strongly biased against the current data.

All things being equal, a system is correctable if there exists some additional information M' such that the sum of $M + M'$ results in an amount of information sufficient for predicting correctly:

Definition 7.10 (Correctable System) Given $\Gamma^W \mid\sim^{trained} \psi, \xi$ and $f(W) = M$, then $C(S, \psi) < 1$ if and only if $\exists M'$ such that $f(W') = M + M' = N$ and $\Delta^{test, W'} \vdash^{ground} \neg\psi, \xi'$.

Note that so far we have presented the correction distance as a distance measure from an incorrect to a correct prediction. Nonetheless, using the correction distance as a criterion to establish whether a classifier is biased against a certain class essentially abstracts away from such a direction (incorrect to correct). In other words, one can ignore the assumption of initial incorrectness of the system and ask what would take for the system to change (or flip, in a binary setting) the classification of a given datapoint, and whether that amount of information is different for distinct classes of individuals.

To reach a proper definition of bias, we need to accommodate the last desideratum left unaddressed so far, i.e., the idea of skewness as "partiality of the domain affected by the error" introduced in Sect. 7.3.2 at point (2). Put differently, while the correction distance has been defined for an individual classification $\psi := T(a)$, now it must be generalised over a subdomain of $\mathbb{D}^{test}$ in order to fully account for the notion of machine learning bias we want to capture:

Definition 7.11 (Measure of Group Fairness)

$$Fairness_{group}(S, T) = \mid C(S, T(a)) - C(S, T(b)) \mid < \epsilon$$

$$T \in \mathbb{T}, \forall a, b \in \mathbb{D}, Q \in \mathbb{PR} \mid Q(a) \wedge \neg Q(b)$$

and where ϵ is a threshold value for the difference of the correction measures above which fairness is considered to fail.

Fairness is defined in terms of parity (liberalised through a tolerance threshold ϵ) in the correction distances across the subdomains of $\mathbb{D}^{test}$ individuated by a binary protected feature Q occurring in the features set Γ^W which defines S.

Coming to the measure of individual fairness, we explore a strategy inspired by the principle of Fairness Through Unawareness (or Blindness) mentioned in Sect. 7.2.1. To implement it, we adopt the following similarity criterion. In general, given two sets A and B, the Jaccard Index of A and B, denoted by $J(A, B)$, is defined as:

Definition 7.12 (Jaccard Index)

$$J(A, B) = \frac{\mid A \cap B \mid}{\mid A \cup B \mid}$$

Two individuals a and b in the test sample $\mathbb{D}^{test}$, given the sets of the predicates they respectively satisfy (denoted by $\mathbb{L}_a$ and $\mathbb{L}_b$ with slight abuse of notation), are called "blindly similar"—expressed by $BSim(a, b)$—if and only if (1) they only differ with respect to the protected predicate Q, and (2) they show more similarities than differences:

Definition 7.13 (Blind Similarity) Two datapoints a and b are blindly similar, denoted by $BSim(a, b)$, iff:

1. $(\mathbb{L}_a \cup \mathbb{L}_b) \setminus (\mathbb{L}_a \cap \mathbb{L}_b) = \{Q\} \in \mathbb{PR}$
2. $J(\mathbb{L}_a, \mathbb{L}_b) > \frac{1}{2}$

Definition 7.14 (Measure of Individual Fairness)

$$Fairness_{Ind}(S, T) = \mid C(S, T(a)) - C(S, T(b)) \mid < \epsilon$$

where

$$T \in \mathbb{T}, \forall a, b \in \mathbb{D} \mid BSim(a, b)$$

and where ϵ is a threshold value for the difference of the correction measures above which fairness is considered to fail.

Namely, we define a measure for individual fairness in terms of parity (liberalised through a tolerance threshold ϵ) in the correction distances of similar individuals, according to the definition of similarity established by Definition 7.13.

The two fairness measures just given (Definitions 7.11 and 7.14) reflect the distinct sizes of weighted information sets required by each group (respectively, individual) to obtain a new, correct classification, given that an incorrect one has been given by the classifier. Note that these measures will reflect the measure of the difference $\xi' - \xi$ as it occurs in the conclusion of the $correction$ rule. Furthermore, if we intend the $correction$ rule as a step-by-step application that only adds one element to the context at each instance (and implements Weakening, until it fails), this measure will correspond to the length of such derivation.

7.6 Conclusions and Further Work

In this work, we proposed a novel logical understanding and associated formalisation of machine learning bias. Based on an underlying notion of bias *qua* prediction error of a peculiar type (implicitly and widely accepted, at least across the wide literature on statistical measures of fairness Verma and Rubin 2018), we pinpointed four key desiderata for a logic capable of modelling the behaviour of a biased system. Namely: skewness, dependency on data and model, non-monotonicity, and the existence of a minimal distinction between types of bias.

After that, we developed a logical framework implementing all these requirements. We structured our proposal at two complementary levels. At a syntactical level, we modelled biased predictions in terms of a non-monotonic and probabilistic inference relation. Within this deduction system, the notions of feature importance, ground truth, training and test sample were used to model correction process of a prediction.

From the point of view of semantics, we modelled the way a biased agent produces an incorrect outcome, and provided a distance measure from its correction in terms of accuracy of the system, quantity of information available, and quantity of information sufficient for correctly predicting the input. On this basis, we defined two metrics, one for group and one for individual fairness.

The main goal for the future development of this work is the developing of a sound semantics for this inference relation, with associated model checking and correctness algorithms to formally verify fairness. To this end, it can be useful to use a modal setting: Kripke models can be interpreted as sets of feature attributions to datapoints, and a modality can be defined to probabilistically evaluate the satisfaction of new features, including the target one. Under this interpretation, individual fairness, as formalised in Definition 7.14, could be verified as the safety property that, if two worlds are similar (according to Definition 7.13), so are the paths to worlds they access such that they satisfy the target feature with the same probability.

Acknowledgments This research was supported by the Ministero dell'Università e della Ricerca (MUR) through PRIN Projects BRIO—Bias, Risk and Opacity in AI (2020SSKZ7R), SMART EST—Simulation of Probabilistic Systems for the Age of the Digital Twin (2022E8Y4X) and through the Project "Departments of Excellence 2023-2027" awarded to the Department of Philosophy "Piero Martinetti" of the University of Milan.

References

Albini, E., S. Sharma, S. Mishra, D. Dervovic, and D. Magazzeni. 2023. On the Connection Between Game-Theoretic Feature Attributions and Counterfactual Explanations. In *Proceedings of the 2023 AAAI/ACM Conference on AI, Ethics, and Society, AIES '23, New York, NY, USA, 2023*, 411–431. Association for Computing Machinery.

Belle, V. 2023. Toward a Logical Theory of Fairness and Bias. *Theory and Practice of Logic Programming* 23 (4): 865–883.

Buolamwini, J., and T. Gebru. 2018. Gender Shades: Intersectional Accuracy Disparities in Commercial Gender Classification. In *Conference on Fairness, Accountability and Transparency, FAT 2018, 23-24 February 2018, New York, NY, USA*. Proceedings of Machine Learning Research, vol. 81, 77–91. PMLR.

Calegari, R., and F. Sabbatini. 2022. The Psyke Technology for Trustworthy Artificial Intelligence. In *AIxIA 2022 - Advances in Artificial Intelligence - XXIst International Conference of the Italian Association for Artificial Intelligence, AIxIA 2022, Udine, Italy, November 28 - December 2, 2022, Proceedings*, ed. Agostino Dovier, Angelo Montanari, and Andrea Orlandini. Lecture Notes in Computer Science, vol. 13796, 3–16. Springer.

Calegari, R., G. Ciatto, and A. Omicini. 2020. On the Integration of Symbolic and Sub-Symbolic Techniques for XAI: A Survey. *Intelligenza Artificiale* 14 (1): 7–32.

Castelnovo, A., R. Crupi, G. Greco, and D. Regoli. 2021. The Zoo of Fairness Metrics in Machine Learning. CoRR, abs/2106.00467.

Chouldechova, A., and A. Roth. 2018. The Frontiers of Fairness in Machine Learning. arXiv preprint arXiv:1810.08810.

Coraglia, G., F.A. D'Asaro, F.A. Genco, D. Giannuzzi, D. Posillipo, G. Primiero, and C. Quaggio. 2023. BRIOxAlkemy: A Bias Detecting Tool. In *Proceedings of the 2nd Workshop on Bias, Ethical AI, Explainability and the role of Logic and Logic Programming co-located with the 22nd International Conference of the Italian Association for Artificial Intelligence (AI*IA 2023)*, ed. Guido Boella, Fabio Aurelio D'Asaro, Abeer Dyoub, Laura Gorrieri, Francesca A. Lisi, and Giuseppe Primiero Chiara Manganini. CEUR Workshop Proceedings, vol. 3615, 44–60. CEUR-WS.org.

D'Asaro, F.A., and G. Primiero. 2021. Probabilistic Typed Natural Deduction for Trustworthy Computations. In *Proceedings of the 22nd International Workshop on Trust in Agent Societies (TRUST2021 @ AAMAS)*.

D'Asaro, F.A., F. Genco, and G. Primiero. 2023. Checking Trustworthiness of Probabilistic Computations in a Typed Natural Deduction System. arXiv preprint arXiv:2206.12934.

De-Arteaga, M., A. Dubrawski, and A. Chouldechova. 2018. Learning Under Selective Labels in the Presence of Expert Consistency. CoRR, abs/1807.00905.

Eubanks, V. 2018. *Automating Inequality*. New York: St. Martin's Press.

Fogliato, R., A. Chouldechova, and M. G'Sell. 2020. Fairness Evaluation in Presence of Biased Noisy Labels. In *International Conference on Artificial Intelligence and Statistics*, 2325–2336. PMLR.

Gajane, P. 2017. On Formalizing Fairness in Prediction with Machine Learning. CoRR, abs/1710.03184.

Genco, F.A., and G. Primiero. 2023. A Typed Lambda-Calculus for Establishing Trust in Probabilistic Programs. CoRR, abs/2302.00958.

Hooker, S. 2021. Moving Beyond "Algorithmic Bias Is a Data Problem. *Patterns* 2: 100241.

Huang, X., and J. Marques-Silva. 2023. The Inadequacy of Shapley Values for Explainability. ArXiv, abs/2302.08160.

Ignatiev, A., M.C. Cooper, M. Siala, E. Hebrard, and J. Marques-Silva. 2020. Towards Formal Fairness in Machine Learning. In *Principles and Practice of Constraint Programming: 26th International Conference, CP 2020, Louvain-La-Neuve, Belgium, September 7–11, 2020, Proceedings*, 846–867. Berlin/Heidelberg: Springer-Verlag.

Kawamoto, Y. 2020. An Epistemic Approach to the Formal Specification of Statistical Machine Learning. *Software and Systems Modeling* 20 (2): 293–310.

Lakkaraju, H., J. Kleinberg, J. Leskovec, J. Ludwig, and S. Mullainathan. 2017. The Selective Labels Problem: Evaluating Algorithmic Predictions in the Presence of Unobservables. In *Proceedings of the 23rd ACM SIGKDD International Conference on Knowledge Discovery and Data Mining, KDD '17*, 275–284. New York: Association for Computing Machinery.

Linardatos, P., V. Papastefanopoulos, and S. Kotsiantis. 2020. Explainable ai: A Review of Machine Learning Interpretability Methods. *Entropy* 23 (1): 18.

Liu, X., and E. Lorini. 2023. A Unified Logical Framework for Explanations in Classifier Systems. *Journal of Logic and Computation* 33 (2): 485–515.

Manganini, C., and G. Primiero. 2023. Reasoning with Bias. In *Proceedings of the 1st Workshop on Fairness and Bias in AI co-located with 26th European Conference on Artificial Intelligence (ECAI 2023), Kraków, Poland, October 1st, 2023*, ed. Roberta Calegari, Andrea Aler Tubella, Gabriel González-Castañé, Virginia Dignum, and Michela Milano. CEUR Workshop Proceedings, vol. 3523. CEUR-WS.org.

Marques-Silva, J. 2023. Disproving xai Myths with Formal Methods—Initial Results. In *2023 27th International Conference on Engineering of Complex Computer Systems (ICECCS)*, 12–21. Los Alamitos: IEEE Computer Society.

Mehrabi, N., F. Morstatter, N. Saxena, K. Lerman, and A. Galstyan. 2021. A Survey on Bias and Fairness in Machine Learning. *ACM computing surveys (CSUR)*, 54 (6): 1–35.

Miceli, M., J. Posada, and T. Yang. 202.. Studying Up Machine Learning Data. *Proceedings of the ACM on Human-Computer Interaction* 6 (GROUP): 1–14.

Olteanu, A., C. Castillo, F. Diaz, E. Kıcıman. 2019. Social Data: Biases, Methodological Pitfalls, and Ethical Boundaries. *Frontiers in Big Data* 2.

Pombal, J., A.F. Cruz, J. Bravo, P. Saleiro, M.A. Figueiredo, and P. Bizarro. 2022. Understanding Unfairness in Fraud Detection Through Model and Data Bias Interactions. arXiv preprint, arXiv:2207.06273.

Primiero, G., and F.A. D'Asaro. 2022. Proof-Checking Bias in Labeling Methods. In *Proceedings of 1st Workshop on Bias, Ethical AI, Explainability and the Role of Logic and Logic Programming (BEWARE 2022) co-located with the 21st International Conference of the Italian Association for Artificial Intelligence (AI*IA 2022), Udine, Italy, December 2, 2022*, ed. Guido Boella, Fabio Aurelio D'Asaro, Abeer Dyoub, and Giuseppe Primiero. CEUR Workshop Proceedings, vol. 3319, 9–19. CEUR-WS.org.

Seshia, S.A., A. Desai, T. Dreossi, D. Fremont, S. Ghosh, E. Kim, S. Shivakumar, M. Vazquez-Chanlatte, and X. Yue. 2018. Formal Specification for Deep Neural Networks. In *Proceedings of the International Symposium on Automated Technology for Verification and Analysis (ATVA)*, 20–34.

Sharma, S., J. Henderson, and J. Ghosh. 2020. Certifai: A Common Framework to Provide Explanations and Analyse the Fairness and Robustness of Black-Box Models. In *Proceedings of the AAAI/ACM Conference on AI, Ethics, and Society, AIES '20*, 166–172. New York: Association for Computing Machinery.

Sharma, S., A.H. Gee, D. Paydarfar, and J. Ghosh. 2021. Fair-n: Fair and Robust Neural Networks for Structured Data. In *Proceedings of the 2021 AAAI/ACM Conference on AI, Ethics, and Society, AIES '21*, 946–955. New York: Association for Computing Machinery.

Suresh, H., and J.V. Guttag. 2019. A Framework for Understanding Sources of Harm Throughout the Machine Learning Life Cycle. *Equity and Access in Algorithms, Mechanisms, and Optimization*. arXiv preprint, arXiv:1901.10002.

Verma, S., and J. Rubin. 2018. Fairness Definitions Explained. In *Proceedings of the International Workshop on Software Fairness, FairWare '18*, 1–7. New York: Association for Computing Machinery.

Chapter 8
Knowledge Representation, Scientific Argumentation and Non-monotonic Logic

Jürgen Landes, Esther Anna Corsi, and Paolo Baldi

Abstract Models developed by the knowledge representation and reasoning community permit us to study defeasible inference based on argumentation and data. Scientific reasoning progresses by evaluating scientific hypotheses based on data and meta-evidence. Meta-evidence can be understood as arguments for discounting or even ignoring data or other meta-evidence. Non-monotonic reasoning is underpinned by non-monotonic logic. We here develop a method for modelling scientific inferences within formal argumentation models. We show how these models capture hypothesis testing, meta-analysis, strong inference and non-monotonic consequence relations.

Keywords Formal argumentation · Non-monotonic logic · Knowledge representation and reasoning · Hypothesis testing · Strong inference · Sally Clark · Rényi-Ulam game · Meta-evidence

8.1 Introduction

The recent years saw rapid advances in data-driven AI and data-driven science more widely. Our capacities to create, store and reason with ever larger data-sets have led to spectacular discoveries and highly successful applications. Yet, most important

J. Landes (✉)
Munich Center for Mathematical Philosophy, LMU Munich, Munich, Germany
e-mail: juergen_landes@yahoo.de

E. A. Corsi
Logic, Uncertainty, Computation and Information (LUCI) Lab, Department of Philosophy, University of Milan, Milan, Italy
e-mail: esther.corsi@unimi.it

P. Baldi
University of Salento, Department of Human Studies, Lecce, Italy
e-mail: paolo.baldi@unisalento.it

H. Hosni, J. Landes (eds.), *Perspectives on Logics for Data-driven Reasoning*, Logic, Argumentation & Reasoning 35,
https://doi.org/10.1007/978-3-031-77892-6_8

issues in science as well as in our daily lives, are not conclusively settled solely based on big data.

Serious debates among sensible people are instead often resolved by *argumentation* since arguments provide (more or less) compelling reasons for siding with one side of the dispute rather than the other. Debates in science are regularly about *scientific hypotheses*, e.g., is a particular hypothesis true, is it useful for applications or what is the hypothesis best supported by the available evidence? *Knowledge representation and reasoning* (KRR) approaches provide models mapping out logical spaces of rational thinking. This chapter develops a formal analysis of scientific debates by drawing on recent advances in one such KRR approach, namely *formal argumentation in AI*.

Argumentation concerns reasons and evidence to support or discard specific claims. Argumentation theory provides a framework for analyzing the strength of arguments and helps individuals to engage in more effective communication. In the context of scientific inference, argumentation theory can play a crucial role in understanding and evaluating the process of forming scientific discoveries.

Scientific hypotheses are an integral part of scientific inference. Once the hypotheses are formed, legions of scientists are interested in determining their status. Paradigmatic procedures for the determination of their status are A) Null Hypothesis Significance Testing (NHST), permitting the rejection of a null hypothesis given data and B) Strong Inference, simultaneously assessing all the hypotheses in a set of hypotheses, given data; in order to refine the possibilities. However, there is now widespread agreement that the data are not to be taken at face value since the data production set-ups and publication processes produce data that does not always track the state of the world. Modern-day scientific inference hence distinguishes between *data* bearing on the status of hypotheses (first-order evidence) and *meta-evidence* concerning the quality of evidence. The challenge then arises, to model the multiple ways in which a piece of meta-evidence interacts with data and further pieces of meta-evidence.

Knowledge Representation and Reasoning is a central and dynamic area of AI that involves the use of formalisms to represent information about the world (Brachman and Levesque 2004; Levesque 1986). These formalisms enable the translation of real-world knowledge into a format that a computer system can understand and manipulate. Once knowledge is represented, the system needs to be able to draw conclusions from that knowledge and such inference procedures involve applying logical rules or other reasoning mechanisms such as probabilistic reasoning, belief revision, argumentation, etc. Despite the rapid progress in statistical techniques and machine learning (ML), KRR remains crucial for certain types of AI applications, especially those that require an explicit representation of knowledge and reasoning about relationships and concepts, such as the relation between an agent and facts of the world (e.g. 'Theo knows φ', 'Theo believes φ', 'Theo fears that φ' etc., where φ stands for a generic proposition that can be true or false). The integration of KRR with ML can help overcome some of the limitations associated with pure ML techniques, particularly in terms of accountability and trustworthiness.

Formal argumentation in AI is a fast-developing field with many successful applications permitting the resolution of conflicting arguments. It is hence well-suited as a formal tool to model intricate relationships between items of evidence in scientific inference and provide a rational way of selecting sets of arguments on the basis of which scientific debates can be settled.

Surprisingly, formal argumentation theory is rarely applied to debates concerning scientific hypotheses, some rare exceptions are Aydinonat et al. (2021), Borg et al. (2019), Cramer and Dauphin (2020), Hahn and Hornikx (2016), Šešelja (2021), Šešelja and Straßer (2013), and Walton and Reed (2005). We hence take steps towards the development of a framework for analysing scientific debates via formal argumentation theory. In doing so we discover a direct connection to *non-monotonic consequence relations*. These have been much studied in the AI community in earlier influential works Kraus et al. (1990) and Makinson (2005), see also, e.g. Marquis et al. (2020) and Strasser and Aldo Antonelli (2019) for a contemporary perspective, and Kyburg and Man Teng (2006) for applications to statistical inference. For formal connections between argumentation frameworks and other formalisms for non-monotonic reasoning, such as default logic, answer set programming, belief revision and non-monotonic consequence relations we refer the reader to Arieli et al. (2019), Heyninck (2019), Heyninck et al. (2023), and Toni and Sergot (2011). The connection with non-monotonic consequence relations allows users of our framework to leverage the work of logicians to better understand rational debates and formal argumentation theory. Links between logic and formal argumentation have received previous attention, e.g., Besnard et al. (2009), Caminada and Gabbay (2009), Efstathiou and Hunter (2011), Gorogiannis and Hunter (2011), Modgil and Prakken (2013), Straßer and Šešelja (2011), and Wu et al. (2009).

The rest of the chapter is organised as follows. Section 8.2 introduces the basics of the currently most popular formal argumentation theory, Dung-Style Argumentation (DSA). Section 8.3 puts forward a distinction between data (Sect. 8.3.1) and meta-evidence (Sect. 8.3.2). We first develop a DSA approach to reasoning with scientific hypotheses based on Dung-Style Argumentation in the absence of meta-evidence. We first model Null Hypothesis Significance Testing (Sect. 8.3.1.1), meta-analysis (Sect. 8.3.1.2) and Strong Inference (Sect. 8.3.1.3) in the absence of meta-evidence. In Sect. 8.3.2, we generalise our model to include meta-evidence. Section 8.4 offers some conclusions.

8.2 Dung-Style Argumentation Theory

8.2.1 Background

In this chapter, we focus on Dung-style Argumentation Theory, originally introduced by Dung (1995), which is currently widely used in AI. Part of the appeal of the framework is surely its simplicity and its malleability. Drastically oversimplifying matters, argumentation uses directed graphs, with vertices and arrows, as a general framework for representing non-monotonic reasoning. Intuitively, an argument that an epistemic agent may endorse is represented by a node, and an attack between arguments by a directed edge between nodes. For any such directed graph, the semantics compute sets of good (jointly acceptable) arguments that can be seen as reasonable points of view. Such a setting is *nonmonotonic* or *defeasible*, since it allows that, in view of additional arguments or attacks, an agent may retract her conclusions and remove some arguments from her accepted set of arguments.

This is a very simplistic picture of formal argumentation abstracting argumentation practices. While this simplicity over-simplifies in a number of respects, the underlying formalism lends itself to a number of modifications that allow for much richer inference procedures that better capture our argumentative practices. For example, we can easily adapt the model to represent (a) supporting arguments (adding a "support" relation) (Cayrol and Lagasquie-Schiex 2005), (b) multiple arguments jointly attacking different arguments (adding a hyper-graph relation) (Nielsen et al. 2007) and (c) strengths of attacks and defences (turning binary attack/support relations into quantitative ones) (Amgoud et al. 2017).

We can also consider argumentation as a multi-agent undertaking in which agents provide parts of the information and might not know all the information provided by other agents, e.g., Tohmé et al. (2008). There have also been competitions to design AI agents that obtain argumentative goals when facing other (adversary) agents (Cerutti et al. 2019), see also Herzig and Yuste Ginel (2021) and Landes and Buettner (2012).

A third important stream of research has focused on developing and assessing different ways of picking out sets of arguments (semantics) given an argumentative context. Argumentative semantics can be defined in two ways: using extensions or labels. The labelling approach is more general than the extension-based approach, but without a doubt, the latter is more intuitive and easier to work with. While the labels can be arbitrary in principle, in practise they tend to include the labels "accepted", "rejected" and "undecided". So, labelling-based semantics are essentially sets of "reasonable" label assignments among all possible ones. Extensions-based semantics can be viewed as compatible sets of arguments that can withstand external attacks. See Baroni et al. (2011) for an extensive analysis on argumentative semantics labelling-based and extension-based.

8.2.2 *The Formal Framework*

An **argumentation framework** (AF) (Dung 1995) is a pair $\mathcal{A} = \langle Ar, R_{\rightarrow} \rangle$ where:

1. Ar is a set of arguments, and
2. $R_{\rightarrow}$ is a binary relation over Ar ($R_{\rightarrow} \subseteq Ar \times Ar$).

Whenever A and B are two arguments in Ar, $(A, B) \in R_{\rightarrow}$ means that the argument A attacks the argument B. In the sequel, instead of writing $(A, B) \in R_{\rightarrow}$, we will use the more intuitive notation $A \longrightarrow B$ and also use $\longrightarrow$ to denote $R_{\rightarrow}$ itself.

The question arises which arguments one may jointly hold at any point in time. The minimal requirement is taken to be that the set of arguments one currently holds should be *conflict-free*. That is, no currently held argument can attack another such argument. The next important property is that the held arguments can *defend* themselves: the set of currently held arguments S must attack every other argument that attacks an argument in S. See Table 8.1 for the most popular semantics and their definitions. Note that a single argumentation semantics may have one (possibly empty) or multiple extensions when applied to a fixed argumentation framework. Their relationships are given in Fig. 8.1 and some of their important properties in

Table 8.1 Definitions of popular extension-based semantics

Name	Definition of semantics S
Admissible	Conflict-free and defends all its elements
Complete	Admissible and contains every argument it defends
Grounded	Minimal (w.r.t. set inclusion) complete extension
Preferred	Maximal (w.r.t. set inclusion) admissible extension
Stable	Conflict-free and S attacks every a in $Ar \setminus$ S
Ideal	Maximal (w.r.t. set inclusion) admissible set That is contained within every preferred extension

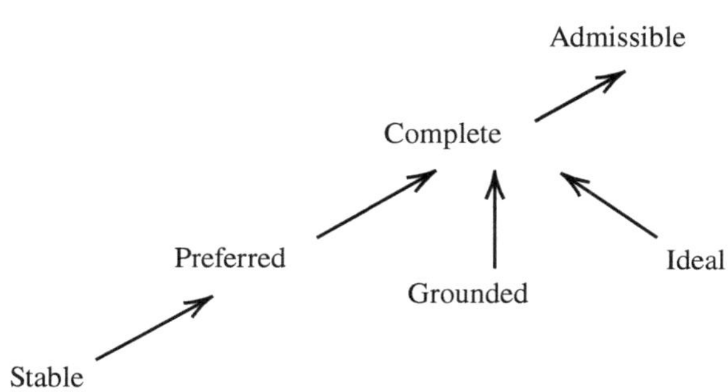

Fig. 8.1 Relations of semantics: an arrow indicates sub-set inclusion of semantics, e.g., every stable semantics is also a preferred semantics

Table 8.2 Important properties of popular semantics

Semantics	Property
Admissible	Universally defined (∅ is always admissible)
Complete	Universally defined
Grounded	Unique, universally defined
Preferred	Universally defined
Stable	Not universally defined
Ideal	Unique, universally defined

Table 8.2. The interested reader can refer to Table 2 in Baroni et al. (2011) for a much more complete analysis of the semantics' properties.

8.3 Argumentation Models for Scientific Inference

In this chapter, we distinguish different types of scientific inference concerning the status of scientific hypotheses. The types we describe are idealised to a degree and only cover some scientific inferences. Although they only partially cover our domain of interest, we still believe that the area they cover is sufficiently large to warrant our attention.

We are interested in regular science as opposed to rare eureka moments that dramatically change the frame of reference. We take it that these artistic-seeming moments of brilliance are beyond the reach of formal analysis/logic. Furthermore, we consider the stage of scientific inquiry at which the hypotheses have been formed. We hence focus on hypothesis confirmation rather than hypothesis generation, cf. Erasmus et al. (2022).

8.3.1 Letting the Data Speak

The first broad category of inferences concerning scientific hypotheses we have in mind seeks to determine a relationship between data and a hypothesis, which is as much as possible free from human judgement, aiming for a maximal degree of objectivity.

8.3.1.1 Significance Testing

The historically first paradigmatic approach to determine the status between data and a hypothesis was invented by Ronald Aylmer Fisher. Scientific hypotheses tend to posit a (possible causal) relationship between variables. Fisher's main idea was to test the hypothesis that there is no relationships between these variables.

The absence of a relationship is a hypothesis in its own right and called the null hypothesis, H_0.[1]

Experimental data is then used to test H_0. If the observed data are unlikely[2] the null hypothesis is said to be *rejected*. If instead the data are (somewhat) probable under the null, then the null hypothesis cannot be rejected. A failure to reject the null does not mean that the null is/ought to be accepted.

The underlying inference is a variant of Modus Tollens (read $\therefore$ as therefore)

	If H_0 is true, then observed data D is unlikely.
	D was observed.
$\therefore$	H_0 is rejected.

Clearly, one should not infer that H_0 has been dis-proven. Furthermore, one should not infer that the hypothesis of interest has been established. All that has been shown is that observed data are unlikely given that the null hypothesis is true, see Mayo (2018) for much more background on significance testing.

Null Hypothesis Significance Testing

Fisher put forward a "threshold of significance" approach to decide whether observed data is unlikely or not. This threshold, α, represents one's attitude towards the relationship between two error probabilities: (1) rejecting a true null (2) not rejecting a false null.[3]

Although Fisher clearly stated that the choice of α has to be adjusted to circumstances, much of the statistical practice that followed adopted his value of $\alpha = 5\%$. Debates concerning a/the appropriate default p-value threshold have not

---

[1] In practice there are often good grounds for believing the null hypothesis to be false at the stage of extensive evidence gathering. Ethics committees tend to block clinical trials from being conducted when presented with little to no evidence supporting an initial belief in the benefits a medical intervention can provide. Exposing patients in the trial arm to a useless intervention when there are alternative treatment options available is considered unethical. Legal systems only allow trials to take place, if the prosecution can provide some evidence that links a putative perpetrator to a crime.

[2] In some more but not full detail, the data being unlikely means that under the assumption of the null hypothesis, the p-value, i.e. the likelihood of observing (the mean of the) data at least as extreme as the observed data is below some threshold. Instead, simply requiring that data is unlikely will result in spurious rejections of the null, since every observed sample consisting of more than a few observations has a low likelihood of obtaining.

[3] Note that, when an experiment is framed in terms of choice between the null hypothesis and an explicitly stated *alternative*, error (2) would amount to failing to accept the alternative hypothesis. In this setting, error (2), which is typically denoted by β cannot be directly calculated from error (1), which is typically still denoted by α.

yet abated (Benjamin et al. 2017). The spread of this method has been so great, that there are now papers written on the misconceptions of p-values (Goodman 2008).

Independently of the proper choice of α, the underlying logical inference naïvely proceeds as follows: test the null hypothesis by performing a number of studies. If the data in one of the studies has a p-value below α, then the null hypothesis is rejected. There is a logical analogy to the legal concept of proof beyond reasonable doubt: The innocence of a defendant (null hypothesis) is rejected by a single line of evidence establishing guilt beyond a threshold of reasonable doubt.

Model

Our model captures this naïve inference procedure as follows applying Dung-style Argumentation Theory. The null hypothesis is an argument as are all the relevant studies concerning H_0, which we denote by $D_1, \ldots, D_n$, the data. Endogenously given a value of α, we have that a D_i attacks H_0, if and only if the p-value of D_i is less than α. Typically, when further *structure* is added to an argumentation framework (Besnard et al. 2014; Heyninck 2019), the attack (and support) relations carry with them either (a) logical relationships (the attacking argument's conclusion logically negating an element of the attacked argument) or (b) reference to background knowledge. Instead, attacks are here defined in terms of conditional probabilities. Note that since data cannot attack other data and the null can't attack data, there are no further attacks in the argumentation framework. We then define the obvious semantics by letting H_0 be rejected, if and only if at least one D_i attacks H_0. See Fig. 8.2 for an example of this simplistic model. While the model is surely basic it is highly malleable and will be modified below.

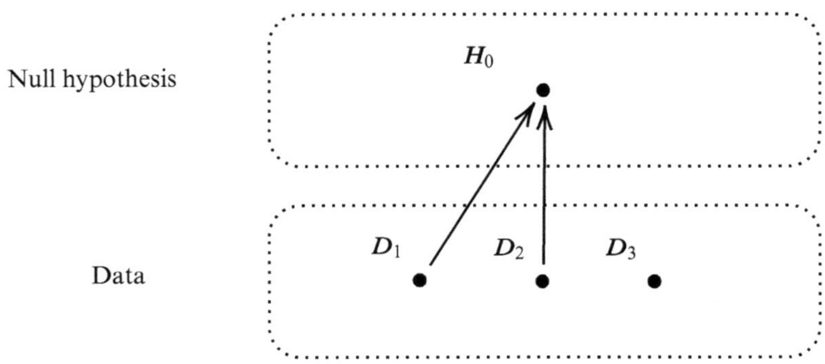

Fig. 8.2 Schema for an argumentation framework for NHST in which D_1, D_2 attack H_0. The null is hence rejected

Example: Legal and Mathematical Reasoning
Although our focus is on data-intensive science, it is worth noting that this version of our model captures legal and mathematical reasoning. We can interpret H_0 as the hypothesis that a defendant is innocent, the D_i as reliable and independent witnesses (or lines of evidence) and the attack relation between a D_i and H_0 as establishing guilt.

Mathematical and many physical theories are taken to have a universal domain applying at all times and in all places. A single case, in which the theory is confirmed to fail, is enough to warrant the rejection of the entire theory. To be clear, it is not necessarily clear what a better theory is; all we can conclude is that a better theory must accommodate the case bringing down the original theory. We here interpret H_0 as a mathematical or physical theory, the D_i are deductive consequences of H_0 and an attack between a D_i and H_0 is present as soon as we learn that a D_i fails to hold. Clearly, the underlying logic is classical.

8.3.1.2 Evidence Aggregation

While it is appropriate to consider the mathematical consequences of a hypothesis separately, data pertaining to scientific statistical hypotheses is often better assessed jointly. That is, the question is not whether the D_i are on their own unlikely given the null but rather whether the set of all data is unlikely given the null.

Meta-Analysis

The statistical methodology to jointly assess the (un-)likelihood of multiple studies is known as meta-analysis. While meta-analysis is an active area of research (Borenstein et al. 2009; Vovk and Wang 2020), the basic idea is to test whether all the data have a joint p-value less than a specified α. If so, then the meta-analysis is said to reject the null. Otherwise, the null is retained. One key methodological advantage of meta-analysis is that it can aggregate all the observations from a number of studies and thus potentially drastically reduce random error.

Model
We can capture this inferential procedure by merging all D_i into a single argument D and let D attack the null, if and only if the p-value of the meta-analysis is below a given threshold α. See Fig. 8.3 for an illustration.

8.3.1.3 Multiple Hypotheses

One natural critique of null hypothesis significance testing is that there is no alternative hypothesis on offer in case the null is rejected. The alternative suggestion

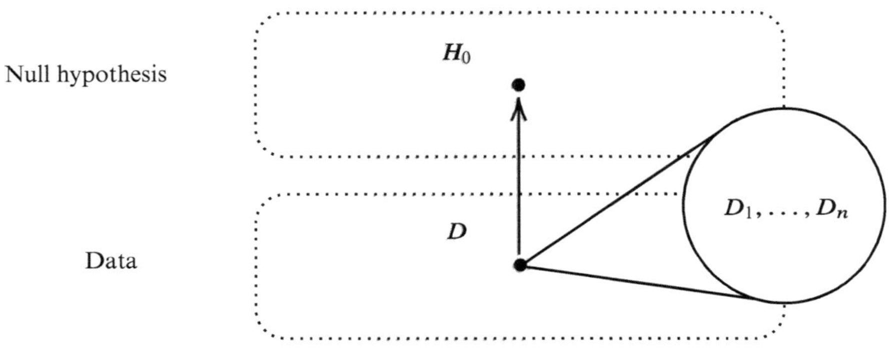

Fig. 8.3 Schema for an argumentation framework meta-analysis

is to consider a set of mutually exclusive hypotheses, $\mathcal{H} = \{H_1, \ldots, H_k\}$, and determine the hypothesis, which is currently best-supported by the available evidence.

Strong Inference

In a 1964 *Science* editorial, John Platt put forward a methodology for scientific inference with multiple hypotheses dubbed *Strong Inference* (Platt 1964) following this schema:

1. Hypothesis Generation.
2. Devising Experiment(s).
3. Experimentation.
4. Go back to 1.

Step 3 permits scientists to discard hypotheses based on the outcome of experiments.

Applying this inferential algorithm one devises ever more detailed hypotheses, which we judge by devising and carrying out ever more crucial experiments that allow us to exclude more and more hypotheses. Progressing through this "logical tree" is in Platt's words a "regular method for reaching firm inductive conclusions one after the other as rapidly as possible".

Model

Our model now contains arguments $H_1, \ldots, H_k$ representing the set of hypotheses and arguments $D_1, \ldots, D_n$. Depending on the preferred relationship between data and hypotheses, our model takes a different form.

(A) In the *NHST tradition*, we can say that a D_i attacks H_j, if and only if the p-value of D_i given H_j is below some threshold of significance α.

(B) Following the *meta-analytic tradition* we can say that all D_i merged into a single argument D attack an H_j, if and only if D is unlikely given H_j. We can then define semantics in the obvious way: H_j is rejected, if and only if D attacks H_j.

(C) In a *Rényi-Ulam game*, a player seeks to determine the true hypothesis H_s among $H_1, \ldots, H_k$ by asking another player, that we here call Nature, questions, that can only be answered by either 'Yes' or 'No'. Nature is allowed to lie at most $m \geq 0$ times. For every fixed $m \in \mathbb{N}$ it provides sound and complete semantics for $m + 2$-valued Łukasiewicz logic. In the borderline case of $m = 0$ the logic collapses to classical logic. In a scientific setting, experiments map into questions of the game and data into answers provided by Nature. Taking inspiration from the Rényi-Ulam game, we thus introduce a semantics for our argumentation framework, labelling the H_j as follows:

rejected, if H_j is attacked by at least $m + 1$ arguments D_i;
least attacked, if every other H_k is attacked at least as often as H_j,
accepted, if it is not rejected and every other hypothesis is rejected and
undecided, otherwise.

Non-Monotonic Logic

The labelling presented above can be used to define a consequence relation, where the collected data take the role of the premises and the hypotheses the place of conclusion(s). The idea is that the data entail, according to such consequence relation, the disjunction of *all* the hypotheses that are *least attacked* by the data, and the classical logical consequences of this disjunction. For instance, if $\mathcal{H} = \{H_1, H_2, H_3, H_4\}$ and the collected data D_1, D_2 attack H_1, H_2, respectively, then one can infer $H_3 \vee H_4$, and logical consequences thereof.

Clearly, as more data is collected, we are not guaranteed that the hypotheses which were previously least attacked will remain least attacked. This means that, contrary to the *monotonicity* of classical logic, according to such a consequence relation, as the set of premises grows, the set of justifiable conclusions may actually shrink. The consequence is thus part of the family of *non-monotonic logics*.

8.3.2 Meta-Evidence

The available data (dis-)confirming scientific hypotheses simultaneously depends on the true state of the world and the data production set-ups. Unfortunately, not all data production set-ups generate data that is free of biases (systematic error). The scale of biases in published scientific papers shocked many when the replication crisis erupted finding that more than 80% of experiments failed to replicate in some fields (Begley and Ioannidis 2015). While the replication crisis did not start scientific research into the production of scientific evidence, it has certainly fuelled its growth (Ioannidis et al. 2015). Consequently, bodies of evidence concerning important scientific hypotheses nowadays consist of two parts: the primary data obtained directly pertaining to the scientific hypotheses, and evidence concerning the production of scientific evidence. Above, we have referred to the former as *data* (first-order evidence) and will call the latter as *meta-evidence*. Note that the production of meta-evidence is also subject to research. Hence, a piece of meta-evidence may also be informative about another piece of meta-evidence, and there is thus no need for higher orders of meta-evidence. Those interested in a philosophical discussion on the nature of evidence may consult (Fletcher 2022; Williamson 2015).

Scientific hypotheses and the body of meta-evidence are largely independent in the sense that the latter does not concern the scientific hypotheses under consideration themselves but rather the data (Friedman and Šešelja 2023; Lasonen-Aarnio 2014). A number of tools have been proven useful for generating meta-evidence such as funnel plots, GRADE (GRADE Working Group 2004; Guyatt et al. 2008), the Catalogue of Bias (Thomas and Heneghan 2022) and ROBINS-I (Sterne et al. 2016).

Unfortunately, the problem of how to quantify the influence of the meta-evidence on confirmation remains an unsolved conundrum; see Casini and Landes (2024) for a Bayesian treatment. The multitude of tools available does not simplify matters but instead exacerbates the problem of quantification (Stegenga 2014). The relationship between scientific hypotheses and meta-evidence in current scientific practice is thus complicated and fails to be "objective".

8.3.2.1 Reasoning with Meta-Evidence

The presence and weight of meta-evidence forces us to leave behind the idea that we can rationally take the available data at face value. Instead, we have to painstakingly examine if and if so to which degree meta-evidence bears on data and on meta-evidence. See the following example.

Example: Meat and Cancer
The *Annals of Internal Medicine* published a number of papers claiming that the data purportedly evidencing the claim that "meat causes cancer" was "weak and based on low certainty evidence". This is an instance of meta-evidence attacking data. The body of meta-evidence was later argued to itself be the result of low quality research methodology (Neuhouser 2019; Potter and Jackson 2019), see National Academies of Sciences (2023, pages 33 and 24) and De Pretis et al. (2022) and Jukola (2019) for further discussion. The evidence for this later arguments is an example of meta-

(continued)

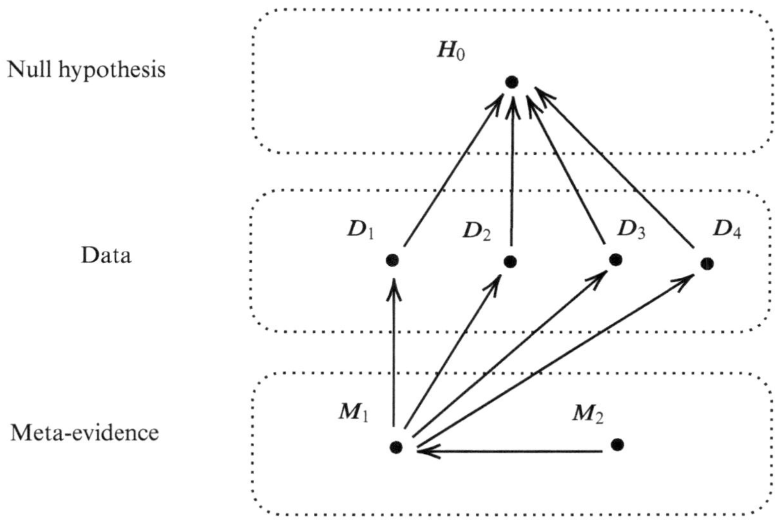

Fig. 8.4 In the absence of meta-evidence, the null is rejected. Introducing M_1 and the attacks originating from M_1 discredits the data, and the null is retained. Also including M_2 and the attacks originating from it, undermines the credibility of M_1, thus the D_i are re-instated, which in turn leads to a rejection of the null

evidence pertaining only to other meta-evidence but not the data nor to the hypothesis under investigation. See Fig. 8.4 for a graphical representation and further explanations.

In an effort to keep matters simple, one may choose to base the status of a scientific hypothesis/hypotheses on all data, which has not been discredited by the available meta-evidence.

8.3.2.2 Model

We again model every hypothesis of interest (either H_0 or $H_1, \ldots, H_k$) by an argument. Similarly, all studies directly pertaining to the hypothesis/hypotheses continue to be modelled by arguments. The attack relation between a single D_i and a hypothesis remains also unchanged. Every item of meta-evidence is represented by an argument, denoted by $M_1, \ldots, M_l$. An M_j can attack an argument D_i, representing a claim of deficient data production. An M_j can also attack another meta-evidence argument M_o, which represents a claim of a deficient item of meta-evidence. We intend M_j to attack M_o to be interpreted as M_j saying that M_o is deficient. Since meta-evidence concerns other items of evidence but does not bear

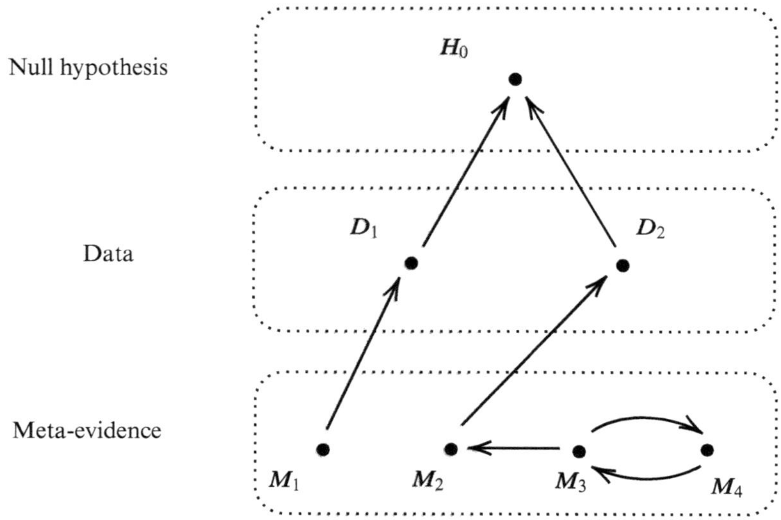

Fig. 8.5 Example argumentation framework for NHST with data and meta-evidence

directly on the scientific hypothesis/hypotheses of interest, the M_j do *not attack* hypotheses. By contrast, the D_i may attack hypotheses.

The set of arguments of the arising AF can be naturally divided into three layers. The top (first) layer contains either the null hypothesis H_0 or the set of competing hypotheses $\mathcal{H} = \{H_1, \ldots, H_k\}$. The middle (second) layer contains the data, $D_1, \ldots, D_n$. The bottom (lower) layer contains the meta-evidence, $M_1, \ldots, M_l$. Only three types of attack may occur: from the second layer to the first, from the third layer to the second and from the third layer to the third layer, see Fig. 8.5 for an example containing all three types of attack.

The D_i on which we want to base our assessment of the status of a hypothesis are those which are not discredited by the meta-evidence. We thus employ a semantics and apply it to the AF. In order to carry out these assessments we insist on using a semantics which only has a single extension $S \subseteq Ar$ when employed to the AF, see Sect. 8.3.2.3 for further comments. The key observation is that the application of the semantics puts us back into the situations we covered in Sect. 8.3.

Null Hypothesis Significance Testing

In the case of NHST, our model for meta-evidence has an extension that consists of a subset of the D_i. If at least one of the D_i in this subset attacks the null hypothesis H_0, then H_0 is rejected. If no such D_i attacks the null hypothesis H_0, then H_0 is retained. We thus are back in the situation we discussed in Sect. 8.3.1.1.

Example: The Sally Clark Case
In 1999, Sally Clark, a 35-year-old mother, was convicted of murdering her two newborn children. During the trial, Professor Meadow, a paediatrician and expert witness, claimed that the chance of one child dying from Sudden Infant Death Syndrome (SIDS) given that the child has not been murdered (the presumption of innocence) was 1 in 8500, and consequently, the probability of having two children dying from SIDS was twice that: 1 over 73 million—the evidence. Considering this probability as too low, this evidence alone was sufficient for Mrs Clark's conviction. In 2003, the case was reopened and new evidence was presented in court that refuted the probability calculated by Professor Meadow. The conviction was overturned and Sally Clark was set free. Let us denote by H_0 hypothesis that Mrs Clark is innocent, and with D_1 the evidence used by Professor Meadow led to the defendant's conviction. The arguments leading to the rejection of D_1 can be summarised as follows:

M_1: *Assumption of Independence*. Meadow wrongly assumed that the death of the second child was independent of the death of the first child.
M_2: *Unreliable Witness*. Professor Meadow is not a Statistician.
M_3: *Prosecutor's fallacy*. The probability of having two infants dying of SIDS, assuming the innocence of Mrs Clark, has been confused with the probability of the innocence of Mrs Clark, given the evidence. In this specific case, the evidence observed is the deaths of the two children and two alternative explanations of the evidence should have been considered: SIDS or murder. Thus, it should have been computed the relative value of the probability of having two children die from SIDS conditional on the evidence against, still conditional on the evidence, the probability of the two children being murdered. It can be shown that, given plausible assumptions, this ratio is equal to 3 (Dawid 2003). Therefore, it is three times more likely that the two kids died because of SIDS instead of being murdered.
M_4: *Ecological fallacy*. The characteristics of one part should not be confused with the characteristics of a whole.Thus, the probability of SIDS in Mrs Clark's family may not be the same as the probability calculated for a group of families with similar characteristics (and used in D_1 by Prof. Meadow).

See Fig. 8.6 for a graphical representation of the argumentation framework that formalises the Sally Clark case.

The underlying logical structure of the Sally Clark case is that of a murder trial. It is non-monotonic since adding meta-evidence can invert the normatively correct judgement, but, at the same time, it has in common with classical logic the fact that

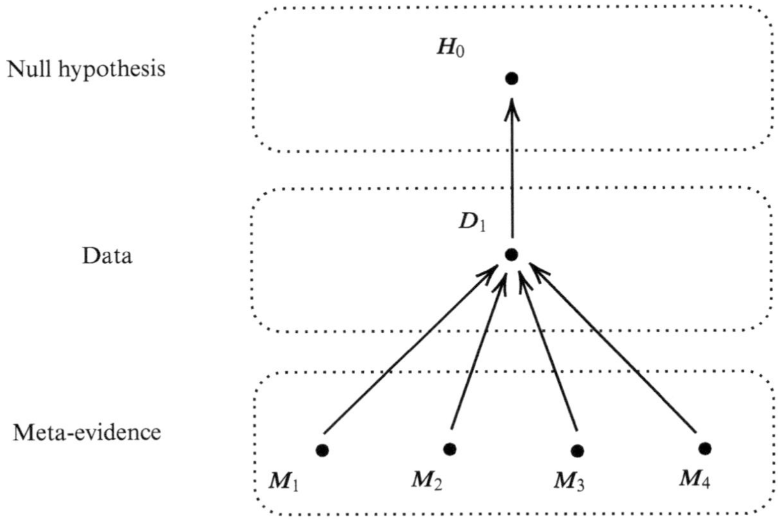

Fig. 8.6 Argumentation framework for the Sally Clark case

a single sufficiently credible line of evidence suffices for the rejection of the null hypothesis (innocence of the defendant).

For legal reasoning and other theories with analogous behaviour it does not matter in how many instances the theory turned out to be confirmed/how many lines of independent evidence putatively establishing the guilt of the defendant the defence attorneys managed to discredit, one single case sinks the theory/establishes guilt. We are here writing with a broad brush. More intricate inferences such as the following are outside the scope of this chapter: The defence managed to show that all but one line of evidence presented by prosecution were flawed, therefore the belief in the entire evidence production methodology has eroded and the only other line of evidence linking the defendant to the crime is thus also tarnished, cf. Collins et al. (2018) for assessing the reliability of sources.

Example: An Expanding Universe

Before the work of Einstein on the general theory of relativity in 1915 (Einstein 1915), cosmology, the science of the origin and development of the universe, was not even considered science, and the scientific community considered the universe as an entity not subject to change, without spacial limits and without temporal limits (hypothesis H_0). In 1922, Friedmann, building on Einstein's works, published a paper Friedman (1922) proposing the first theory of an expanding universe. The work of Friedmann can be seen

(continued)

as first-order evidence (D_1) that attacks H_0. After a few months, Einstein published a letter Einstein (1922) in the same journal as Friedmann's paper, pointing out some calculation errors (meta-evidence M_1). Subsequently, Einstein acknowledged Friedman's calculations were correct and retracted his letter Einstein (1923) (meta-evidence M_2). Lemaître's theory on the expansion of the universe, which also was a consequence of Einstein's work, was published in 1927 (Lemaître 1927) (evidence D_2). Einstein, while finding no formal errors in Lemaître's work, called it "abominable" and remained sceptical about the possibility of the universe expanding (Friedmann and Lemaître 1997) (argument M_3). Although going against Einstein's intuition might seem foolish, it is crucial to base scientific claims on evidence. Therefore, it can be argued that Einstein's criticism of Lemaître's work lacks a solid foundation. For this reason, we introduce the argument M_4 that attacks the argument M_3. In 1929, Hubble's experimental evidence (Hubble 1929) definitively confirmed Lemaître and Friedmann's theories.

Even though the arguments D_1 and D_2 do not consist of data, they should be considered as first-order evidence. In fact, they are not the product of abstruse mental lucubrations but are logical[4] consequences of existing theories. For this reason, D_1 and D_2 should not be understood simply as the hypotheses scientists can formulate, but they consist of evidence (theoretical evidence) that attacks H_0. Argument D_3 differs from D_1 and D_1 in nature and presents experimental evidence rather than theoretical. See Fig. 8.7 for a schematic representation of the scientific debate just recalled concerning the status of the universe.

H_0: The universe is fixed and without limit in both time and space.

D_1: Work of Friedmann.
D_2: Work of Lemaître.
D_3: Hubble's data.

M_1: Computational errors in Friedmann's work.
M_2: Friedmann's calculations are correct.
M_3: Einstein's scepticism towards Lemaître's theory.
M_4: Scientific criticism has to be based on evidence, not intuition.

[4] Here we refer to classical logic.

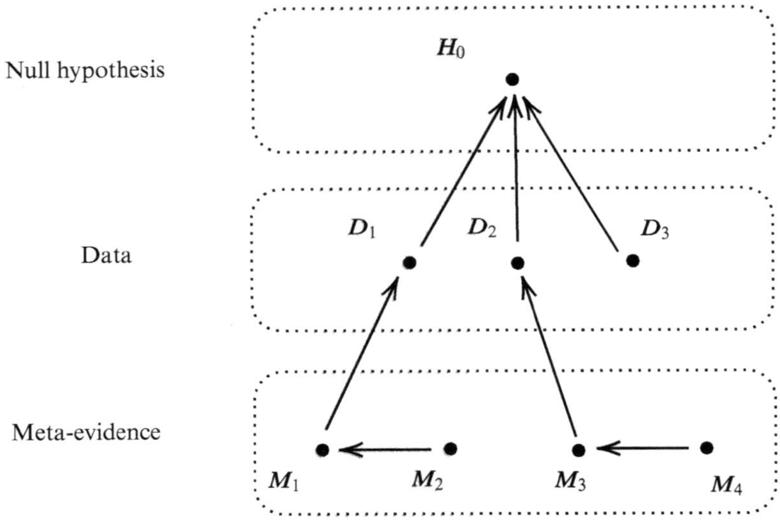

Fig. 8.7 Representation of the scientific debate that led at the beginning of the twentieth century to the discarding of the idea that the universe is fixed and unchanging

In the context of physics, a given hypothesis H_0 can be rejected using theory or data. However, as shown in the example above, these types of attacks do not carry equal weight. Finding theoretical evidence that attacks H_0 is not a sufficient condition for rejecting it, and it is only with the empirical evidence that supports theoretical attacks that H_0 can be rejected. In this context, theoretical and empirical evidence could also be considered as two parts of a single argument.

Evidence Aggregation

In case the data are to be aggregated jointly, a meta-analysis is performed by aggregating all the primary data picked out by the extension S, as explained in Sect. 8.3.1.2. Whether the data in the meta-analysis is sufficiently implausible to reject the null hypothesis cannot be determined by the information contained in the AF. Recall instead that in the case of assessing the null hypothesis based on multiple studies separately, the information contained in the AF suffices to determine the status of the null hypothesis.

Multiple Hypotheses

We again use the extension S to pick out a set of arguments, the D_i. If there are multiple hypotheses on the table, two possibilities arise. Firstly, a hypothesis H_j is assessed based on all D_i in the extension S separately. Secondly, every hypothesis is assessed based on all D_i in the extension S jointly. Either way, we continue as explained in Sect. 8.3.1.3.

Non-Monotonic Logic

By the same idea described in Sect. 8.3.1.3, our argumentation model determines a non-monotonic consequence relation: *acceptable* pieces of data entail the disjunction of all the hypotheses that are least attacked by the data, and its classical logical consequences. The main difference now is that taking into account meta-evidence, we assess hypotheses only on the basis of acceptable pieces of data, i.e. those that belong to the unique extension identified by the semantics of the underlying argumentation model.

8.3.2.3 Levels of Generality

Our approach can be generalised in a number of directions that we shall now briefly discuss and explain why we will not follow up on them here.

1. We insisted on the extension S being a subset of Ar rather than a set of subsets of Ar. In case multiple subsets of Ar are deemed reasonable by the semantics, defining the set of least attacked hypotheses is not straightforward. For example, define a degree of attack $d(\mathbb{D}, H)$ as the number of attacks from D to H. One may then define a hypothesis to be the least attacked if $\inf_{\mathbb{D}\in S} d(\mathbb{D}, H)$ is minimal. However, one could instead replace the $\inf_{\mathbb{D}\in S}$ by $\sup_{\mathbb{D}\in S}$ or average $d(\mathbb{D}, H)$ over all $\mathbb{D} \in S$. Alternatively, one may take the union of all $\mathbb{D} \in S$ or the intersection of all $\mathbb{D} \in S$ and determine "least attacked" based on the arguments in this set of arguments. Since the science methodology literature has not addressed this issue, we shall refrain from proposing a formal approach here.
2. We did not consider graded attacks from meta evidence on other evidence. Our main reason for doing so is that there are no widely accepted semantics for this generalisation, although some suggestions have been made (Grossi and Modgil 2019; Hunter et al. 2021).
3. We did not allow hypotheses to attack other hypotheses. In case the hypotheses are mutually exclusive, one is naturally inclined to think that a hypothesis attacks every other hypothesis. While this is so, it does not add useful information to the argumentation framework and one may as well drop all these attacks. In case the hypotheses are not mutually exclusive, some attacks among hypotheses may be present. The challenge then arises to amalgamate the attacks from the data with the attacks from other hypotheses. Since the science methodology literature has not suggested any measure, we have abstained from putting forward an ill-motivated measure of amalgamation.
4. We have treated all hypotheses as equals. If there is pertinent further information available, some sort of Bayesian approach incorporating prior beliefs in hypotheses may be added to our picture. Since it is our intention to be maximally neutral on statistical disputes, we have refrained from using Bayesian methodology.

8.4 Conclusions

In this chapter, we put forward a way to model scientific inference with tools developed in formal argumentation theory situated within the Knowledge Representation and Reasoning field.

Limitations of our approach are that (1) it fails to clearly link to a decision after a status of a/the hypothesis/hypotheses has been established. (2) It requires categorical judgements and does not permit more nuanced assessments, e.g., either an attack is present or absent.

In order to address the second point a graded notion of argumentation is required, which permits attacks to be gradual. A logical next step to take is to add uncertainty and/or attacks that come with a degree formalising the idea that some attacks are more powerful than others. There's more than one way in which uncertainties can be introduced; some of which are reviewed in Hunter et al. (2021). The obvious alternative seems to go fully Bayesian (Landes et al. 2018), which also takes care of the first limitation by clearly linking to rational decision-making permitting the maximisation of expected utilities. This, however, runs counter to our starting point of being true to scientific practice and hold on to, as much as possible, an objective (data-driven) notion of probability.[5]

We believe that our framework has a number of appealing features that invite fruitful discussions, applications, adaptations and later improvements.

1. It faithfully represents regular scientific practice to a large degree.
2. It covers a sizeable part of scientific debate and inference.
3. It makes explicit the distinction between data and meta-evidence and highlights their different inferential roles.
4. It captures the intuition that (1) there is an objective relationship between evidence and hypotheses and at the same time (2) a subjective element of evidence assessment is required.
5. It opens up the grounds of (non-)monotonic logics for applications to improve our understanding of science.
6. It is made up of distinguishable parts capturing various aspects of scientific reasoning. This property permits straightforward substitutions of parts of the framework by (1) other models more appropriate in other contexts and (2) more advanced models as new modelling techniques and approaches are developed.

In recent work Baldi et al. (2024, forthcoming), we presented non-monotonic consequence relations based on the notion of *degrees of rejection*, where data entail the classical consequences of all the hypotheses which are minimally rejected. Inspired by that, here we sketched a consequence relation, which is derived on a principled manner from argumentation frameworks. Importing formal concepts from this latter field, we were thus able to improve on Baldi et al. (2024,

[5] Another option is to adopt a different statistical paradigm as suggested in Mayo and Hand (2022).

forthcoming): in particular, while Baldi et al. (2024, forthcoming) only dealt with evidence and hypotheses, we have here incorporated into the consequence relation also an account of the meta-evidence, using suitable argumentation semantics.

Future work, besides overcoming the above-mentioned limitations, may be conducted in a number of further directions. (1) Show how fallacious reasoning can be captured as a violation of at least one of our assumptions/axioms. (2) Model further inferences such as those in tort law. There, the crucial status of a hypothesis is that 'a hypothesis has a rational probability of at least 50% on the preponderance of the evidence'. (3) A more ambitious project is to seek to determine the rational semantics (singular and/or plural) used in science to deal with the available meta-evidence by developing a (number of) case-study.

Acknowledgments We thank Matthias Thimm for his helpful comments in getting us started on this chapter.

Funding Jürgen Landes gratefully acknowledges NextGenerationEU funding for the project "Practical Reasoning for Human-Centred Artificial Intelligence" as well as funding from the Deutsche Forschungsgemeinschaft (DFG, German Research Foundation)—project number 528031869. Paolo Baldi acknowledges support from the project "Ethical Design for AI", part of the Spoke 6 of the PNRR project FAIR (PE00000013, CUP H97G22000210007), funded by the European Union-NextGenerationEU

References

Amgoud, L., J. Ben-Naim, D. Doder, S. and Vesic, S. 2017. Acceptability Semantics for Weighted Argumentation Frameworks. In: *Proceedings of IJCAI, pp. 56–62. International Joint Conferences on Artificial Intelligence Organization*, ed. C. Sierra. https://doi.org/10.24963/ijcai.2017/9.

Arieli, O., A. Borg, J. and Heyninck. 2019. A Review of the Relations Between Logical Argumentation and Reasoning with Maximal Consistency. *Annals of Mathematics and Artificial Intelligence* 87 (3): 187–226. https://doi.org/10.1007/s10472-019-09629-7.

Aydinonat, N.E., S. Reijula, and P. Ylikoski, P. 2021. Argumentative Landscapes: The Function of Models in Social Epistemology. Synthese 199 (1–2): 369–395. https://doi.org/10.1007/s11229-020-02661-9.

Baldi, P., E.A. Corsi, and H. Hosni. 2024, forthcoming. A Logical Framework for Data-Driven Reasoning. *Logic Journal of the IGLP*. https://doi.org/10.1093/jigpal/jzae113.

Baroni, P., M. Caminada, and M. Giacomin 2011. An Introduction to Argumentation Semantics. *Knowledge Engineering Review* 26 (4): 365–410. https://doi.org/10.1017/S0269888911000166.

Begley, C.G., and J.P. Ioannidis. 2015. Reproducibility in Science. *Circulation Research* 116 (1): 116–126. https://doi.org./10.1161/CIRCRESAHA.114.303819.

Benjamin, D.J., J.O. Berger, M. Johannesson, B.A. Nosek, E.J. Wagenmakers, R. Berk, K.A. Bollen, B. Brembs, L. Brown, C. Camerer, et al. 2017. Redefine Statistical Significance. *Nature Human Behaviour* 2: 6–10. https://doi.org/10.1038/s41562-017-0189-z

Besnard, P., A. Hunter, and S. Woltran. 2009. Encoding Deductive Argumentation in Quantified Boolean Formulae. *Artificial Intelligence* 173 (15): 1406–1423. https://doi.org/10.1016/j.artint.2009.06.006.

Besnard, P., A. Garcia, A. Hunter, S. Modgil, H. Prakken, G. Simari, and F. Toni. 2014. Introduction to Structured Argumentation. *Argument and Computation* 5 (1): 1–4. https://doi.org/10.1080/19462166.2013.869764.

Borenstein, M., L.V. Hedges, J.P. Higgins, and H.R. Rothstein. 2009. *Introduction to Meta-Analysis*. New York: Wiley.

Borg, A., D. Frey, D. Šešelja, and C. Straßer. 2019. Theory-Choice, Transient Diversity and the Efficiency of Scientific Inquiry. *European Journal for Philosophy of Science* 9 (2). https://doi.org/10.1007/s13194-019-0249-5.

Brachman, R., and H. Levesque. 2004. *Knowledge Representation and Reasoning*. San Francisco: Morgan Kaufmann.

Caminada, M., and D. Gabbay. 2009. A Logical Account of Formal Argumentation. *Studia Logica* 93: 109–145. https://doi.org/10.1007/s11225-009-9218-x.

Casini, L., and J. Landes. 2024. Bias, Conflict of Interest, and the Principle of Total Evidence. In *Proceedings of Reasoning with Imperfect Information in Social Settings*, ed. M. Piazza, M. Tesi, and P. Vigani, 24 pp.

Cayrol, C., and M.C. Lagasquie-Schiex. 2005. On the Acceptability of Arguments in Bipolar Argumentation Frameworks. In *Proceedings of ECSQARU, LNCS*, vol. 3571, 378–389. Springer. https://doi.org/10.1007/11518655_33.

Cerutti, F., M. Giacomin, and M. Vallati. 2019. How We Designed Winning Algorithms for Abstract Argumentation and Which Insight We Attained. *Artificial Intelligence* 276: 1–40. https://doi.org/10.1016/j.artint.2019.08.001.

Collins, P.J., U. Hahn, Y. von Gerber, and E.J. Olsson. 2018. The Bi-Directional Relationship Between Source Characteristics and Message Content. *Frontiers in Psychology* 9: 1–16. https://doi.org/10.3389/fpsyg.2018.00018.

Cramer, M., and J. Dauphin. 2020. A Structured Argumentation Framework for Modeling Debates in the Formal Sciences. *Journal for General Philosophy of Science* 51 (2): 219–241. https://doi.org/10.1007/s10838-019-09443-z.

Dawid A. 2003. *Sally Clark Appeal: Statement of Professor A. P. Dawid*. http://www.statslab.cam.ac.uk/~apd/SallyClark_report.doc.

De Pretis, F., S. Jukola, and J. Landes. 2021. E-Synthesis for Carcinogenicity Assessments: A Case Study of Processed Meat. *Journal of Evaluation in Clinical Practice* 28: 752–772. https://doi.org/10.1111/jep.13697.

Dung, P.M. 1995. On the Acceptability of Arguments and Its Fundamental Role in Nonmonotonic Reasoning, Logic Programming and N-Person Games. *Artificial Intelligence* 77 (2): 321–357. https://doi.org/10.1016/0004-3702(94)00041-X.

Efstathiou, V., and A. Hunter. 2011. Algorithms for Generating Arguments and Counterarguments in Propositional Logic. *International Journal of Approximate Reasoning* 52 (6): 672–704. https://doi.org/10.1016/j.ijar.2011.01.005.

Einstein, A. 1915. Die feldgleichungen der gravitation Sitzungsberichte der Königlich Preußischen Akademie der Wissenschaften, 844–847.

Einstein, A. 1922. Bemerkung zu der Arbeit von A. Friedmann "Über die Krümmung des Raumes". *Zeitschrift für Physik* 11: 326–326.

Einstein, A. 1923. Notiz zu der Arbeit von A. Friedmann "Über die Krümmung des Raumes". *Zeitschrift für Physik* 16: 228–228.

Erasmus, A., B. Holman, and J.P.A. Ioannidis. 2022. Data-Dredging Bias. *BMJ Evidence-Based Medicine* 27 (4): 209–211. https://doi.org/10.1136/bmjebm-2020-111584.

Fletcher, A.N. 2022. What Is Evidence as Evidence Is Used? A Case of Dualism? *Social Theory & Health* 20 (3): 291–305. https://doi.org/10.1057/s41285-021-00170-4.

Friedman, A. 1922. Über die krümmung des raumes. *Zeitschrift für Physik* 10 (1): 377–386.

Friedmann, A., and G. Lemaître. 1997. Essais de cosmologie. Précédé de : Invention du Big Bang, de Jean-Pierre Luminet Seuil.

Friedman, D.C., and D. Šešelja. 2023. Scientific Disagreements, Fast Science and Higher-Order Evidence. *Philosophy of Science*. https://doi.org/10.1017/psa.2023.83.

Goodman, S. 2008. A Dirty Dozen: Twelve *p*-Value Misconceptions. *Seminars in Hematology* 45 (3): 135–140. https://doi.org/10.1053/j.seminhematol.2008.04.003.

Gorogiannis, N., and A. Hunter. 2011. Instantiating Abstract Argumentation with Classical Logic Arguments: Postulates and Properties. *Artificial Intelligence* 175 (9–10): 1479–1497. https://doi.org/10.1016/j.artint.2010.12.003.

GRADE Working Group. 2004. Grading Quality of Evidence and Strength of Recommendations. *British Medical Journal* 328 (7454): 1490–1494. https://doi.org/10.1136/bmj.328.7454.1490.

Grossi, D., and S. Modgil. 2019. On the Graded Acceptability of Arguments in Abstract and Instantiated Argumentation. *Artificial Intelligence* 275: 138–173. https://doi.org/10.1016/j.artint.2019.05.001.

Guyatt, G.H., A.D. Oxman, G.E. Vist, R. Kunz, Y. Falck-Ytter, P. Alonso-Coello, and H.J. Schünemann. 2008. GRADE: An Emerging Consensus on Rating Quality of Evidence and Strength of Recommendations. *British Medical Journal* 336 (7650): 924–926. https://doi.org/10.1136/bmj.39489.470347.AD.

Hahn, U., and J. Hornikx. 2016. A Normative Framework for Argument Quality: Argumentation Schemes with a Bayesian Foundation. *Synthese* 193 (6): 1833–1873. https://doi.org/10.1007/s11229-015-0815-0.

Herzig, A., and A. Yuste Ginel. 2021. Multi-Agent Abstract Argumentation Frameworks With Incomplete Knowledge of Attacks. In *Proceedings of International Joint Conferences on Artificial Intelligence Organization, IJCAI*, ed. Z.H. Zhou, 1922–1928. https://doi.org/10.24963/ijcai.2021/265.

Heyninck, J. 2019. Relations Between Assumption-Based Approaches in Non-monotonic Logic and Formal Argumentation: From Structured Argumentation to Adaptive Logics. *Journal of Applied Logics* 6: 317–357.

Heyninck, J., G. Kern-Isberner, T. Meyer, J.P. Haldimann, and C. Beierle. 2023. Conditional Syntax Splitting for Non-monotonic Inference Operators, 6416–6424. Association for the Advancement of Artificial Intelligence (AAAI). https://doi.org/10.1609/aaai.v37i5.25789.

Hubble, E. 1929. A Relation Between Distance and Radial Velocity Among Extra-Galactic Nebulae. *Proceedings of the National Academy of Sciences* 15 (3): 168–173.

Hunter, A., A. Polberg, N. Potyka, T. Rienstra, and M. Thimm. 2021. Probabilistic Argumentation: A Survey. In *Handbook of Formal Argumentation*, ed. D. Gobbay, M. Giacomin, G.R. Simari, and M. Thimm, chap. 7, 397–444. College Publications.

Ioannidis, J.P.A., D. Fanelli, D.D. Dunne, and S. N. Goodman. 2015. Meta-Research: Evaluation and Improvement of Research Methods and Practices. *PLOS Biology* 13 (10): e1002264. https://doi.org/10.1371/journal.pbio.1002264.

Jukola, S. 2019. On the Evidentiary Standards for Nutrition Advice. *Studies in History and Philosophy of Science Part C: Studies in History and Philosophy of Biological and Biomedical Sciences* 73: 1–9. https://doi.org/10.1016/j.shpsc.2018.05.007.

Kraus, S., D. Lehmann, and M. Magidor. 1990. Nonmonotonic Reasoning, Preferential Models and Cumulative Logics. *Artificial Intelligence* 44 (1–2): 167–207. https://doi.org/10.1016/0004-3702(90)90101-5.

Kyburg, H.E., and C. Man Teng. 2006. Nonmonotonic Logic and Statistical Inference. *Computational Intelligence* 22 (1): 26–51.

Landes, J., and R. Buettner. 2012. Argumentation-Based Negotiation? Negotiation-Based Argumentation! In *Proceedings of EC-Web. LNBIP*, ed. P.L. Christian Huemer, vol. 123, 149–162. Heidelberg: Springer. https://doi.org/10.1007/978-3-642-32273-0_13.

Landes, J., B. Osimani, and R. Poellinger. 2018. Epistemology of Causal Inference in Pharmacology. *European Journal for Philosophy of Science* 8: 3–49. https://doi.org/10.1007/s13194-017-0169-1.

Lasonen-Aarnio, M. 2014. Higher-Order Evidence and the Limits of Defeat. *Philosophy and Phenomenological Research*, 88 (2): 314–345. https://doi.org/10.1111/phpr.12090.

Lemaître, G. 1927. Un Univers homogène de masse constante et de rayon croissant rendant compte de la vitesse radiale des nébuleuses extra-galactiques. *Annales de la Société Scientifique de Bruxelles* 47: 49–59.

Levesque, H.J.: Knowledge representation and reasoning. Annual Review of Computer Science 1(1), 255–287 (1986). https://doi.org/10.1146/annurev.cs.01.060186.001351

Makinson, D. 2005. *Bridges from Classical to Nonmonotonic Logic*. Cambridge: Cambridge University Press.

Marquis, P., O. Papini, and H. Prade. 2020. *A Guided Tour of Artificial Intelligence Research 1: Knowledge Representation, Reasoning and Learning*. Berlin: Springer.

Mayo, D.G. 2018. *Statistical Inference as Severe Testing: How to Get Beyond the Statistics Wars*. Cambridge: Cambridge University Press.

Mayo, D.G., and D. Hand. 2022. Statistical Significance and Its Critics: Practicing Damaging Science, or Damaging Scientific Practice? *Synthese* 200 (3). https://doi.org/10.1007/s11229-022-03692-0.

Modgil, S., and H. Prakken. 2013. A General Account of Argumentation with Preferences. *Artificial Intelligence*, 195: 361–397. https://doi.org/10.1016/j.artint.2012.10.008, corrigendum available at https://doi.org/10.1016/j.artint.2018.05.001.

Neuhouser, M.L. 2019. Red and Processed Meat: More with Less? *American Journal of Clinical Nutrition* 111 (2): 252–255. https://doi.org/10.1093/ajcn/nqz294.

Nielsen, S., and S. Parsons. 2007. A Generalization of Dung's Abstract Framework for Argumentation: Arguing with Sets of Attacking Arguments. In *Proceedings of ArgMAS, LNCS*, ed. N. Maudet, S. Parsons, and I. Rahwan, vol. 4766, 54–73. Springer. https://doi.org/10.1007/978-3-540-75526-5_4.

Platt, J.R. 1964. Strong Inference. *Science* 146 (3642): 347–353. https://doi.org/10.1126/science.146.3642.347.

Potter, J.D., and R. Jackson. 2019. On Meat, Butter, and Fudge. *Nutrition and Cancer* 72 (1): 1–4. https://doi.org/10.1080/01635581.2019.1703837.

National Academies of Sciences Engineering, and Medicine. 2023. *Sponsor Influences on the Quality and Independence of Health Research*. Washington: National Academies Press. https://doi.org/10.17226/27056.

Šešelja, D. 2021. Some Lessons from Simulations of Scientific Disagreements. *Synthese* 198 (S25): 6143–6158. https://doi.org/10.1007/s11229-019-02182-0.

Šešelja, D., and C. Straßer. 2013. Abstract Argumentation and Explanation Applied to Scientific Debates. *Synthese* 190 (12): 2195–2217. https://doi.org/10.1007/s11229-011-9964-y.

Stegenga, J. 2014. Down with the Hierarchies. *Topoi* 33 (2): 313–322. https://doi.org/10.1007/s11245-013-9189-4.

Sterne, J.A., M.A., Hernán, B.C. Reeves, J. Savović, N.D. Berkman, M. Viswanathan, D. Henry, D.G. Altman, M.T. Ansari, I. Boutron, J.R. Carpenter, A.W. Chan, R. Churchill, J.J. Deeks, A. Hróbjartsson, J. Kirkham, P. Jüni, Y.K. Loke, T.D. Pigott, C.R. Ramsay, D. Regidor, H.R. Rothstein, L. Sandhu, P.L. Santaguida, H.J. Schünemann, B. Shea, I. Shrier, P. Tugwell, L. Turner, J.C. Valentine, H. Waddington, E. Waters, G.A. Wells, P.F. Whiting, and J.P. Higgins. 2016. ROBINS-I: A Tool for Assessing Risk of Bias in Non-randomised Studies of Interventions. *BMJ* 355: 1–7. https://doi.org/10.1136/bmj.i4919.

Strasser, C., and G. Aldo Antonelli. 2019. Non-monotonic Logic. In *Stanford Encyclopedia of Philosophy*, ed. Edward N. Zalta, Summer 2019 ed. Metaphysics Research Lab, Stanford University. https://plato.stanford.edu/archives/sum2019/entries/logic-nonmonotonic/.

Straßer, C., and D. Šešelja. 2011. Towards the Proof-theoretic Unification of Dung's Argumentation Framework: an Adaptive Logic Approach. *Journal of Logic and Computation* 21 (2): 133–156. https://doi.org/10.1093/logcom/exq015.

Thomas, E.T., and C. Heneghan. Catalogue of Bias: Selective Outcome Reporting Bias. *BMJ Evidence-Based Medicine* 27: 370–372. https://doi.org/10.1136/bmjebm-2021-111845.

Tohmé, F., G. Bodanza, and G. Simari. 2008. Aggregation of Attack Relations: A Social-Choice Theoretical Analysis of Defeasibility Criteria. In *Proceedings of FoIKS. LNCS*, ed. S. Hartmann, and G. Kern-Isberner, vol. 4932, 8–23. Springer. https://doi.org/10.1007/978-3-540-77684-0_4.

Toni, F., and M. Sergot. 2011.*Argumentation and Answer Set Programming*, 164–180. Berlin: Springer. https://doi.org/10.1007/978-3-642-20832-4_11.

Vovk, V., and R. Wang. 2020. Combining p-Values via Averaging. *Biometrika* 107 (4): 791–808. https://doi.org/10.1093/biomet/asaa027.

Walton, D., and C.A. Reed. 2005. Argumentation Schemes and Enthymemes. *Synthese* 145 (3): 339–370. https://doi.org/10.1007/s11229-005-6198-x.

Williamson, J. 2015. Deliberation, Judgement and the Nature of Evidence. *Economics and Philosophy* 31: 27–65. https://doi.org/10.1017/S026626711400039X.

Wu, Y., M. Caminada, and D. Gabbay. 2009. Complete Extensions in Argumentation Coincide with 3-Valued Stable Models in Logic Programming. *Studia Logica* 93: 383–403. https://doi.org/10.1007/s11225-009-9210-5.

Chapter 9
Reasoning with Data in the Framework of a Quantum Approach to Machine Learning

Maria Luisa Dalla Chiara, Roberto Giuntini, and Giuseppe Sergioli

Abstract *How are abstract concepts formed and recognized on the basis of a previous experience?* It is interesting to compare the behavior of human minds and of artificial intelligences with respect to this problem. Generally, a human mind that abstracts a concept (say, *table*), from a given set of known examples creates a *table-Gestalt*: a kind of vague and out of focus image that does not fully correspond to a particular table with well determined features. Can the construction of a *gestaltic pattern*, which is so natural for human minds, be taught to an intelligent machine? This problem can be successfully discussed in the framework of a quantum approach to machine learning. The basic idea is replacing *classical datasets* with *quantum datasets* where objects are described by special quantum states, involving the characteristic uncertainty and ambiguity of the quantum theoretic formalism. In this framework, the intuitive concept of *Gestalt* can be simulated by the mathematical concept of *positive centroid of a given quantum dataset.* Although recognition procedures are different for human and for artificial intelligences, there is a common method of "facing the problems" that seems to work in both cases. From a logical point if view, a quantum approach to machine learning can be reconstructed in the framework of a special form of quantum-logical semantics that has been suggested by quantum information theory.

M. L. Dalla Chiara
Dipartimento di Lettere e Filosofia, Università di Firenze, Firenze, Italy
e-mail: marialuisa.dallachiara@unifi.it

R. Giuntini
Dipartimento di Pedagogia, Psicologia, Filosofia, Università di Cagliari, Cagliari, Italy

Institute for Advanced Studies, Technical University Munich, Garching, Germany
e-mail: giuntini@unica.it

G. Sergioli (✉)
Dipartimento di Pedagogia, Psicologia, Filosofia, Università di Cagliari, Cagliari, Italy
e-mail: giuseppe.sergioli@unica.it

H. Hosni, J. Landes (eds.), *Perspectives on Logics for Data-driven Reasoning*,
Logic, Argumentation & Reasoning 35,
https://doi.org/10.1007/978-3-031-77892-6_9

Keywords Reasoning with data · Quantum logic · Machine learning · Concept recognition · Quantum datasets

9.1 Introduction

Reasoning with data plays an important role in most researches about pattern recognition and machine learning. A main question that is discussed is: *how are objects classified on the basis of a previous experience?* In many cases, the intuitive concept of *previous experience* is formally represented by means of a convenient notion of *dataset*. In recent times, classification-problems have been successfully investigated in the framework of a *quantum approach to machine learning*, where *objects* are supposed to be described by *pieces of quantum information*: possible states of quantum systems that are storing the information in question. In this way, the characteristic *uncertainties* and *ambiguities* of the quantum-theoretic formalism come into play.

For the readers who are not familiar with quantum mechanics it may be useful to recall some basic concepts of the theory that play an important role in quantum information. Suppose a physicist is studying a quantum physical system **S** (say, an electron) at a given time. His (her) information about **S** can be identified with a particular mathematical object that represents the *state* of **S** at that time. In the happiest situations our physicist might have about **S** a *maximal information* that cannot be consistently extended to a richer knowledge. In such a case, the information in question is called a *pure state of the system*. An observer who has assigned a pure state to a given system knows about this system all that even a hypothetical omniscient mind would know. Following a successful notation introduced by Paul Dirac, quantum pure states are usually denoted by the expressions $|\psi\rangle, |\varphi\rangle, \ldots$ (where $|\ldots\rangle$ represent the so called "ket-brackets"). Mathematically, any pure state $|\psi\rangle$ is a *vector* (with *length* 1) living in a special abstract space, called a *Hilbert space* (usually indicated by the symbol $\mathcal{H}$).[1] A strange logical feature of the quantum formalism is the following: although representing a maximal piece of information, a pure state $|\psi\rangle$ cannot *decide all physical properties* that may hold for a quantum system described by $|\psi\rangle$. Due to Heisenberg's *uncertainty principle* some basic properties (that may concern, for instance, either the position or the velocity) turn out to be essentially *indeterminate*.

In the mathematical formalism of quantum theory *physical properties* are represented as *closed subspaces* of the Hilbert space under consideration (sets of vectors that are closed under linear combinations and limits of Cauchy-sequences). Since the set of all closed subspaces X is in one-to-one correspondence with the set of all projection operators P_X (whose range is the set X), one can equivalently

[1] Hilbert spaces are special examples of vector spaces that represent generalizations of geometric Euclidean spaces. A simple example of a Hilbert space is the geometric *plane*, whose set of *points* corresponds to the set of all possible ordered pairs of real numbers.

assume that *physical properties* are represented by *projection operators* (briefly, *projections*) of the Hilbert space in question.

Given a quantum property represented by the closed subspace X and by the corresponding projection P_X, any pure state $|\psi\rangle$ assigns to X and to P_X a probability-value (indicated by $\mathtt{p}_{|\psi\rangle}(X)$ and by $\mathtt{p}_{|\psi\rangle}(P_X)$), which is determined by the *Born-rule*:

$$\mathtt{p}_{|\psi\rangle}(X) = \mathtt{p}_{|\psi\rangle}(P_X) = \| \; P_X|\psi\rangle \; \|^2 \, .$$

Thus, the probability that a quantum system in state $|\psi\rangle$ verifies the property represented by X and by P_X is the number $\| \; P_X|\psi\rangle \; \|^2$ (the squared length of the vector that has been obtained by projecting $|\psi\rangle$ over the closed subspace X).

Generally, a quantum pure state $|\psi\rangle$ can be represented as a *superposition* (a vector-sum) of other pure states $|\psi_i\rangle$:

$$|\psi\rangle = \sum_i c_i|\psi_i\rangle,$$

where c_i are complex numbers (called *amplitudes*).[2] The physical interpretation of this formal representation is the following: *a quantum system whose state is* $|\psi\rangle$ *might verify the properties that are certain for a system in state* $|\psi_i\rangle$ *with a probability-value that depends on the number* c_i.[3] From an intuitive point of view one can say that a superposition-state seems to describe a kind of *ambiguous cloud of possibilities*: a set of *potential properties* that are, in a sense, all co-existent for a given quantum object.

Not all states of quantum systems are pure. More generally, a piece of quantum information may correspond to a *non-maximal knowledge*: a *mixed state* (or *mixture*), that is mathematically represented as a special Hilbert-space operator called *density operator*. Quantum mixed states give rise to a kind of *second degree of ambiguity*: while any pure state *verifies with certainty* some specific quantum properties, a mixed state may leave indeterminate all non-trivial quantum properties. At the same time, all pure states correspond to special examples of density operators.[4] The Born-rule can be generalized to density operators ρ in the following form (for any projection P):

$$\mathtt{p}_\rho(P) = \mathrm{tr}(\rho P),$$

[2] Complex numbers play a fundamental role in the mathematical representation of many important quantum phenomena (for instance, in the description of *interference-phenomena*).

[3] More precisely, this probability-value is represented by the real number $|c_i|^2$ (the squared modulus of c_i). Since the length of $|\psi\rangle$ is 1, we have: $\sum_i |c_i|^2 = 1$.

[4] Any density operator ρ of a given Hilbert space can be represented (in a non-unique way) as a *weighted sum* of some projections, having the form: $\rho = \sum_i w_i P_{|\psi_i\rangle}$, where the *weights* w_i are positive real numbers such that $\sum_i w_i = 1$, while each $P_{|\psi_i\rangle}$ is the projection that projects over the one-dimensional closed subspace that contains the vector $|\psi_i\rangle$.

Thus, any pure state $|\psi\rangle$ corresponds to a special example of a density operator: the projection $P_{|\psi\rangle}$. The physical interpretation of a mixed state $\rho = \sum_i w_i P_{|\psi_i\rangle}$ is the following: a quantum system in state ρ might be in the pure state $P_{|\psi_i\rangle}$ with probability-value w_i.

where tr is the trace functional. In the case of pure states $|\psi\rangle$ we have:

$$\mathtt{p}_{|\psi\rangle}(X) = \mathrm{tr}(P_{|\psi\rangle}P) = \| P_X|\psi\rangle \|^2 .$$

Complex pieces of quantum information are supposed to be stored by composite quantum systems (say, systems of many electrons). Thus, the quantum theoretic representation of composite systems comes into play, giving rise to one of the most mysterious features of the quantum world: *entanglement*, a phenomenon that had been considered "potentially paradoxical" by some of the founding fathers of quantum theory (for instance, by Einstein and by Schrödinger).

Consider a composite quantum systems

$$\mathbf{S} = \mathbf{S}_1 + \ldots + \mathbf{S}_n.$$

Any state of $\mathbf{S}$ shall live in a particular Hilbert space $\mathcal{H}^{\mathbf{S}}$ that is a special product (called *tensor product*) of the spaces $\mathcal{H}^{\mathbf{S}_1}, \ldots, \mathcal{H}^{\mathbf{S}_n}$, associated to the subsystems $\mathbf{S}_1, \ldots, \mathbf{S}_n$. It is customary to write:

$$\mathcal{H}^{\mathbf{S}} = \mathcal{H}^{\mathbf{S}_1} \otimes \ldots \otimes \mathcal{H}^{\mathbf{S}_n},$$

where $\otimes$ indicates the tensor product.[5]

Unlike the case of classical physics, composite quantum systems have a peculiar *holistic* behavior. The state of the global system $\mathbf{S} = \mathbf{S}_1 + \ldots + \mathbf{S}_n$ determines the states of its component parts $(\mathbf{S}_1, \ldots, \mathbf{S}_n)$, and generally not vice versa. Thus, the procedure goes from the *whole* to the *parts*, and not the other way around.

9.2 Concept-Recognitions for Human and for Artificial Intelligences

How are abstract concepts formed and recognized on the basis of some previous experience?

This difficult question has been investigated with different methods by psychologists, neuroscientists, artificial intelligence researchers, logicians, philosophers. Interestingly enough, neuroscientific researches have recently interacted with an important approach to psychology: the *Gestalt theory*, that had been proposed by Wertheimer, Kofka and Köhler in the early twentieth century.[6] The basic idea of this theory can be sketched as follows: human perception and knowledge of objects is essentially based on our capacity of realizing a *Gestalt* (a *form*) of the objects

[5] The tensor product of two Hilbert spaces $\mathcal{H}_1$ and $\mathcal{H}_2$ is a special Hilbert space that includes (up to isomorphisms) the Cartesian product of $\mathcal{H}_1$ and $\mathcal{H}_2$.

[6] See Wertheimer (1912), Kofka (1935), Köhler (1920) and Ehrenstein et al. (2003).

in question: a *holistic image* that cannot be identified with the set of its component elements. A human mind that abstracts the concept *table* from a given set of concrete examples, generally creates a *table-Gestalt*, a kind of *vague and out of focus image* that does not fully correspond to a particular table with well determined features. When we *recognize* as a *table* a new object we have met in our environment, we generally make a *comparison* between

- the main *features* of the new object;
- the *table-Gestalt* that we had constructed in our mind.

Can such recognition-processes that are so natural for human minds be "taught" to an intelligent machine? Is it possible to *simulate* the intuitive notion of *Gestalt* by some adequate mathematical concepts?

This question can be successfully investigated in the framework of a quantum approach to pattern recognition and to machine learning. Consider an *agent* (let us call her *Alice*) who is interested in a given *concept* C that may refer either to *concrete* or to *abstract* objects (say *table, triangle, beautiful*). The name *Alice* may denote either a *human* or an *artificial intelligence*. We suppose that *Alice* (on the basis of her previous experience) has already *recognized* and *classified* a given set of objects for which the question "does the object under consideration verify the concept C?" can be reasonably asked. We assume that the possible answers to this question are:

- *YES!*
- *NO!*
- *PERHAPS!*

As an example, *Alice* might be a child who has already recognized (in the environment where she is living) the objects that are tables and the objects that are not tables. At the same time, she might have been doubtful about the *right* classification of some particular objects. For instance, she might have answered "PERHAPS!" to the question "is this food-trolley a table?".

While a human *Alice* may have *seen* the objects under consideration, *seeing* is of course more problematic for an *Alice* machine. Thus, generally, one shall make recourse to some *theoretic representation* that faithfully describes the objects in question. As happens in physics, such theoretic representations can be identified with convenient mathematical objects that represent *object-states*. In the classical approaches to machine learning an *object-state* is usually represented as a vector

$$\overrightarrow{\mathbf{x}} = (x_1, \ldots, x_d),$$

that belongs to the real space $\mathbb{R}^d$ (where $d \geq 1$). Every component x_i of the vector $\overrightarrow{\mathbf{x}}$ is supposed to correspond to a possible value of an *observable* that is considered relevant for recognizing the concept C. The number x_i is usually called a *feature* of the object represented by the vector $\overrightarrow{\mathbf{x}}$.

As an example, suppose we are interested in a set of flowers of different kinds (roses,violets, lillies,...). We can assume that each instance of our set is characterized by two features: the petal length and the petal width, represented by two real numbers. Hence the object-state of each flower will be a vector $\overrightarrow{\mathbf{x}} = (x_1, x_2) \in \mathbb{R}^2$.

The main idea of a quantum approach to machine learning is the following: replacing *classical object-states* with *pieces of quantum information*, possible states of quantum systems that are storing the information in question. In many cases it may be useful to start with a classical information represented by a sequence of classical object-states:

$$\overrightarrow{\mathbf{x}_1}, \ldots, \overrightarrow{\mathbf{x}_n},$$

where all vectors $\overrightarrow{\mathbf{x}_i}$ live in the same space $\mathbb{R}^d$. Then, by applying a canonical procedure, each classical state $\overrightarrow{\mathbf{x}_i}$ can be transformed into a corresponding quantum pure state $|\overrightarrow{\mathbf{x}_i}\rangle$ that lives in the (real) Hilbert space $\mathbb{R}^{d+1}$.[7]

We can now discuss in a quantum framework the problem: *how is a concept C recognized on the basis of a previous experience?* Suppose that at a given time *Alice* is interested in the concept C. Her previous experience concerning C can be described by the formal notion of *quantum C-data set.*

Definition 9.1 (Quantum C-dataset) A *quantum C-dataset* is a sequence

$$^{C}DS = (\mathcal{H},\ St,\ St^+,\ St^-,\ St^?\ ),$$

where:

1. $\mathcal{H}$ is a finite-dimensional Hilbert space associated to C.
2. St is a finite set of pure states $|\psi\rangle$ of $\mathcal{H}$ for which the question "does the object described by $|\psi\rangle$ verify the concept C?" can be reasonably asked.
3. St^+ is a subset of St, consisting of all states that have been *positively classified* with respect to the concept C. The elements of this set are called the *positive instances* of the concept C.
4. St^- is a subset of St, consisting of all states that have been *negatively classified* with respect to the concept C. The elements of this set are called the *negative instances* of the concept C.
5. $St^?$ is a (possibly empty) subset of St, consisting of all states that have been considered *problematic* with respect to C. The elements of this set are called the *indeterminate instances* of the concept C.
6. The three sets St^+, St^-, $St^?$ are pairwise disjoint. Furthermore, $St^+ \cup St^- \cup St^? = St$.

We indicate by n, n^+, n^-, $n^?$ the cardinal numbers of the sets St, St^+, St^-, $St^?$, respectively.

[7] Examples of canonical procedures are the *amplitude encoding* and the *stereographic encoding* (see, for instance, Sergioli et al. 2017).

Suppose that at a later time $Alice$ "meets" a new object described by the object-state $|\varphi\rangle$. She shall find a rule that allows her to answer the question "does $|\varphi\rangle$ verify the concept C?". And this answer shall be based on her previous knowledge that is represented by the *quantum C-dataset*

$$^{C}DS = (\mathcal{H},\ St,\ St^{+},\ St^{-},\ St^{?}).$$

A winning strategy is based on the use of two special concepts: the *quantum positive centroid* and the *quantum negative centroid* of a given quantum C-dataset, which are defined in the following way.

Definition 9.2 (Quantum Centroids) Consider a quantum C-dataset

$$^{C}DS = (\mathcal{H},\ St,\ St^{+},\ St^{-},\ St^{?}).$$

(1) The *quantum positive centroid* of ^{C}DS is the following density operator of the space $\mathcal{H}$:

$$\rho^{+} = \sum_i \left\{ \frac{1}{n^{+}} P_{|\psi_i\rangle} : |\psi_i\rangle \in St^{+} \right\}.$$

(2) The *quantum negative centroid* of ^{C}DS is the following density operator of the space $\mathcal{H}$:

$$\rho^{-} = \sum_i \left\{ \frac{1}{n^{-}} P_{|\psi_i\rangle} : |\psi_i\rangle \in St^{-} \right\}.$$

The concept of *quantum positive centroid* seems to represent a "good" mathematical simulation for the intuitive idea of *Gestalt*. In fact, both the *quantum positive centroid* and the intuitive idea of *Gestalt* describe an *imaginary object*, representing a *vague, ambiguous idea* that $Alice$ has obtained as an abstraction from the "real" examples she has met in her previous experience. As happens in the case of the intuitive idea of *Gestalt*, a *quantum positive centroid*, represented by the density operator $\rho^{+} = \sum_i \left\{ \frac{1}{n^{+}} P_{|\psi_i\rangle} : |\psi_i\rangle \in St^{+} \right\}$ *ambiguously alludes* to the concrete *positive instances* that $Alice$ had previously met Giuntini et al. (2023a,b).

We know that *human recognitions and classifications* are usually performed by means of a quick and mostly unconscious *comparison* between the main features of some *new objects* we have met and a *gestaltic pattern* that we had previously constructed in our mind. Any *comparison* generally involves the use of some *similarity-relations* (weak examples of relations that are reflexive, symmetric and generally non-transitive).

Similarity-relations are mostly grasped in a vague and intuitive way by human intelligences. In our quantum approach to machine learning we can use a class of *similarity-relations* defined in terms of an important quantum function, called *fidelity*.

Definition 9.3 (The Fidelity for Pure States) The *fidelity-function* on a Hilbert space $\mathcal{H}$ is defined as the function F that assigns to any pair $|\psi\rangle$ and $|\varphi\rangle$ of pure states of $\mathcal{H}$ the real number

$$F(|\psi\rangle, |\varphi\rangle) = |\langle\psi \mid \varphi\rangle|^2$$

(where $\langle\psi \mid \varphi\rangle$ is the inner product of $|\psi\rangle$ and $|\varphi\rangle$).

From an intuitive point if view, the number $F(|\psi\rangle, |\varphi\rangle)$ can be interpreted as a measure of the *degree of closeness* between the two states $|\psi\rangle$ and $|\varphi\rangle$. The definition of *fidelity* can be generalized to the case of density operators, which may represent either pure or mixed states. Accordingly, in the general case we will write: $F(\rho, \sigma)$.

The concept of *fidelity* allows us to define in any Hilbert space $\mathcal{H}$ a special class of similarity-relations, called *r-similarities*, where r is any real number in the interval [0, 1].

Definition 9.4 (r-Similarity) Let ρ and σ be two density operators of a Hilbert space $\mathcal{H}$ and let $r \in [0, 1]$.
The state ρ is called *r-similar* to the state σ (briefly, $\rho \not\perp_r \sigma$) iff $r \leq F(\rho, \sigma)$.

One can easily check that this relation is reflexive, symmetric and generally non-transitive. When $F(\rho, \sigma) \neq 0$, there are infinitely many r such that $\rho \not\perp_r \sigma$. As we will see, this fact does not represent a shortcoming for our aims.

Now *Alice* has at her disposal the mathematical tools that allow her to face the *classification-problem*. Suppose that *Alice*'s information about a concept C is the *quantum C-dataset*

$$^{C}DS = (\mathcal{H},\ St,\ St^{+},\ St^{-},\ St^{?}),$$

whose *positive* and *negative centroids* are the states ρ^{+} and ρ^{-}, respectively. Let r^* be a *threshold value* in the interval $(\frac{1}{2}, 1]$, that is considered relevant for ^{C}DS. The main goal is defining a *classifier*, that assigns to every state σ (which describes an object that *Alice* may meet) either the value + (corresponding to the answer "YES!") or the value − (corresponding to the answer "NO"!) or the value ? (corresponding to the answer "PERHAPS!").

Definition 9.5 (Classifier) The *classifier*, determined by the quantum dataset ^{C}DS and by the threshold value r^*, is the function $Cl_{[^{C}DS, r^*]}$ that satisfies the following condition for any state σ of the space $^{C}\mathcal{H}$:

$$Cl_{[C_{DS,r^*}]}(\sigma) = \begin{cases} +, & \text{if } \sigma \not\perp_{r^*} \rho^+ \text{ and not } \sigma \not\perp_{r^*} \rho^-; \\ -, & \text{if } \sigma \not\perp_{r^*} \rho^- \text{ and not } \sigma \not\perp_{r^*} \rho^+; \\ ?, & \text{otherwise.} \end{cases}$$

This particular classifier is also called the *fidelity-classifier.*

Example: Fidelity-Classifier

As an example let us consider a simple empirical simulation based on the fidelity-classifier. We suppose that *Alice* is interested in a concept C that describes a kind of flower (say, the *rose*). At the initial time she has a classical information about a (finite) set of different flowers (roses, violets, lilies, ...). Every flower is supposed to be characterized by two features that concern the petal length and the petal width. Thus, the classical object-state that describes each particular flower (in *Alice*'s set of instances) is a vector $\vec{\mathbf{x}_i} = (x_1, x_2)$, that belongs to the real space $\mathbb{R}^2$. Then, *Alice* encodes each classical object-state $\vec{\mathbf{x}_i}$ into a quantum pure state $|\vec{\mathbf{x}_i}\rangle$ of the Hilbert space $\mathbb{R}^3$. On this basis she creates a quantum C-dataset having the following form:

$$^{C}DS = (\mathcal{H},\ St,\ St^+,\ St^-,\ St^?\ ),$$

where C corresponds to the concept *rose*. At a later time, once chosen a threshold-value r^*, *Alice* can apply the fidelity-classifier to any pure state $|\psi\rangle$ (living in the space $\mathbb{R}^3$) that represent the description of a *new* flower that *Alice* may meet.

Although recognition procedures are different for human and for artificial intelligences, there is a common method of "facing the problems" that seems to work in both cases. Using quantum-theoretic concepts represents a great advantage in order to investigate the relationships between the behaviors of human and of artificial intelligences. The intuitive concept of *Gestalt* could hardly been simulated in a classical framework. For, the characteristic *ambiguity* of *quantum positive centroids* is not shared by the corresponding notion of *positive centroid* that is defined in many classical approaches to machine learning: in the classical case a *positive centroid* represents an *exact object-state*, that is obtained by calculating the average-values of the values that all *positive instances* assign to the observables under consideration. Thus, unlike the quantum case, a *classical positive centroid* describes an *imaginary object* that is characterized by *precise features*.[8]

So far we have dealt with quantum datasets that refer to a single concept C. Of course, in the general case, it may be interesting to consider epistemic situations

---

[8] See, for instance, Schuld and Petruccione (2018).

where many different concepts are involved at the same time. These concepts may be either pairwise exclusive (say, *dog, cat, rabbit, ...*) or non-exclusive (say, *black, beautiful, big, ...*). To this aim, we introduce the notion of *quantum dataset-system.*

Definition 9.6 (Quantum Dataset-System) Consider a set of concepts $Conc = \{C_1, \ldots, C_n\}$. A *quantum dataset-system* based on $Conc$ is a system

$$DSSy = \left\{(\mathcal{H}_i,\ St_i,\ St_i^+,\ St_i^-,\ St_i^?)\right\}_{1\leq i\leq n},$$

where every $(\mathcal{H}_i,\ St_i,\ St_i^+,\ St_i^-,\ St_i^?)$ is a C_i-dataset, with $C_i \in Conc$.

Generally, the choice of the space $\mathcal{H}_i$ and of the set St_i depends on the complexity and on the possible ambiguity of the corresponding concept C_i. Notice that one and the same object can be described by different object-states in different datasets of $DSSy$.

9.3 A Quantum Computational Semantics for Quantum Datasets

From a logical point of view quantum dataset-systems can be naturally analyzed in the framework of a special form of quantum semantics, called *quantum computational semantics*, that has been inspired by quantum information theory. The basic intuitive ideas that underlie this semantics can be sketched as follows.

1. *Meanings* of formulas are represented as *pieces of quantum information*: possible states of quantum systems that are storing the information in question.
2. Logical connectives are interpreted as special *reversible* Hilbert-space operators, that describe possible time-evolutions of the quantum system storing a given information. In this way, connectives acquire a *dynamic character*, representing possible computation-actions.

Normally *quantum computational meanings* are supposed to behave in a *holistic* and *contextual* way: the meaning of a molecular formula determines the *contextual meanings* of its well-formed parts (and, generally, not the other way around), against the semantic *compositionality-principle*. Furthermore, meanings may be *context-dependent*: a formula occurring in different contexts may receive different *contextual meanings* in the framework of one and the same model of the language. But in the case of quantum datasets it will be sufficient to consider a simpler form of semantics, called *compositional and non-contextual quantum computational semantics*, where the compositionality-principle is not violated.[9]

[9] See Chiara et al. (2018).

Let us briefly recall the main "mathematical characters" of quantum information, that play an important role in quantum computational semantics. The "mathematical stage" where any piece of quantum information is usually supposed to live is a finite-dimensional Hilbert space whose general form is the following:

$$\mathcal{H}^{(n)} = \underbrace{\mathbb{C}^2 \otimes \ldots \otimes \mathbb{C}^2}_{n-times}$$

(where $n \geq 1$ and $\otimes$ is the tensor product operation).
Accordingly, $\mathcal{H}^{(n)}$ can be regarded as the Hilbert space associated to composite quantum systems consisting of n parts. The number n depends on the complexity of the information stored by the quantum systems in question.

The *canonical orthonormal basis* of the space $\mathcal{H}^{(1)} = \mathbb{C}^2$ consists of the two following pure states:

$$|0\rangle = (1, 0) \text{ and } |1\rangle = (0, 1).$$

In the general case, the *canonical orthonormal basis* of a space $\mathcal{H}^{(n)}$ consists of the pure states $|\psi\rangle$ having the following form:

$$|\psi\rangle = |x_1\rangle \otimes \ldots \otimes |x_n\rangle,$$

where each x_i is either 0 or 1. Instead of $|x_1\rangle \otimes \ldots \otimes |x_n\rangle$ it is customary to write: $|x_1, \ldots, x_n\rangle$.

Any piece of quantum information is a special mathematical object that lives in a particular quantum computational space $\mathcal{H}^{(n)}$, representing a possible state of a physical system that is storing the information in question. Important examples are represented by *quregisters*, *qubits*, *registers*, *bits* and *mixtures of quregisters*.

Definition 9.7 (Quregisters and Registers)

- A *quregister* is a unit vector of a space $\mathcal{H}^{(n)}$.
- A *qubit* is a quregister of the space $\mathbb{C}^2$.
- A *register* is an element of the canonical orthonormal basis of a space $\mathcal{H}^{(n)}$.
- A *bit* is a register of the space $\mathbb{C}^2$.

Quregisters are *pure states* representing *maximal pieces of information* (about the quantum systems under investigation) that cannot be consistently extended to a richer knowledge. Generally, a piece of quantum information may also correspond to a *non-maximal knowledge*: a *mixed state* (a *mixture of quregisters*), mathematically represented as a *density operator* ρ of a space $\mathcal{H}^{(n)}$. Of course, any quregister $|\psi\rangle$ corresponds to a special case of a mixture of quregisters: the projection $P_{|\psi\rangle}$ that projects over the closed subspace determined by $|\psi\rangle$.

Apparently, bits and registers represent classical pieces of information that correspond to a *certain knowledge* in a quantum computational framework. Qubits, quregisters and mixtures of quregisters, instead, represent quantum pieces of information that generally correspond to an uncertain and probabilistic knowledge.

Any quregister $|\psi\rangle$ of a space $\mathcal{H}^{(n)}$ can be represented as a superposition of registers:

$$|\psi\rangle = \sum_i c_i |x_{i_1}, \ldots, x_{i_n}\rangle.$$

As a particular case, any qubit $|\psi\rangle$ (which lives in the space $\mathcal{H}^{(1)} = \mathbb{C}^2$) can be represented as a superposition of the two bits $|0\rangle$ and $|1\rangle$:

$$|\psi\rangle = c_0|0\rangle + c_1|1\rangle.$$

From an intuitive point of view (following the standard interpretation of quantum superpositions) one can say that, generally, a quregister represents a "cloud of possibilities": the uncertain information $|\psi\rangle = \sum_i c_i |x_{i_1}, \ldots, x_{i_n}\rangle$ might be transformed into the certain information $|x_{i_1}, \ldots, x_{i_n}\rangle$ with probability-value $|c_i|^2$. And these virtual possibilities are, in a sense, all co-existent.

As happens in classical information theory, the two bits $|1\rangle$ and $|0\rangle$ can be also regarded as the two classical truth-values 1 and 0. In this way, any qubit $|\psi\rangle = c_0|0\rangle + c_1|1\rangle$ can be logically interpreted as follows: $|\psi\rangle$ represents a (generally) uncertain piece of information that might be *true* with probability $|c_1|^2$ and might be *false* with probability $|c_0|^2$. On this basis, we can consider two special quantum properties represented by the two projections $P_{|1\rangle}$ and $P_{|0\rangle}$. From an intuitive point of view, it seems natural to assume that $P_{|1\rangle}$ and $P_{|0\rangle}$ represent, respectively, the *Truth-property* and the *Falsity-property* for the space $\mathbb{C}^2$. Accordingly, recalling the Born-rule, we obtain that for any qubit $|\psi\rangle = c_0|0\rangle + c_1|1\rangle$:

- $\text{tr}(P_{|\psi\rangle} P_{|1\rangle}) = |c_1|^2$ represents the probability that the piece of information $|\psi\rangle$ is true (i.e. verifies the Truth-property);
- $\text{tr}(P_{|\psi\rangle} P_{|0\rangle}) = |c_0|^2$ represents the probability that the piece of information $|\psi\rangle$ is false (i.e. verifies the Falsity-property).

This procedure can be naturally generalized to any space $\mathcal{H}^{(n)}$. To this aim, let us first distinguish the *true* from the *false* registers of $\mathcal{H}^{(n)}$.

Definition 9.8 (True and False Registers) Let $|x_1, \ldots, x_n\rangle$ be a register of $\mathcal{H}^{(n)}$.

- $|x_1, \ldots, x_n\rangle$ is called *true* iff $x_n = 1$.
- $|x_1, \ldots, x_n\rangle$ is called *false* iff $x_n = 0$.

In other words, the truth-value of any register is determined by its last bit.

On this basis the *Truth-property* and the *Falsity-property* can be defined in any space $\mathcal{H}^{(n)}$ as follows.

Definition 9.9 (Truth and Falsity)

- The *Truth-property* (briefly, the *Truth*) of the space $\mathcal{H}^{(n)}$ is the projection $P_1^{(n)}$ corresponding to the smallest closed subspace that includes the set of all true registers.
- The *Falsity-property* (briefly, the *Falsity*) of the space $\mathcal{H}^{(n)}$ is the projection $P_0^{(n)}$ corresponding to the smallest closed subspace that includes the set of all false registers.

By applying the Born-rule, one can now define a special quantum probability-function $\mathtt{p}$ that assigns to any piece of quantum information ρ a value in the interval $[0, 1]$.

Definition 9.10 (The Probability-Value of a Piece of Quantum Information) For any density operator ρ of a space $\mathcal{H}^{(n)}$,

$$\mathtt{p}(\rho) := \mathrm{tr}(\rho P_1^{(n)}).$$

In the particular case of quregisters $|\psi\rangle$ of a space $\mathcal{H}^{(n)}$, we have:

$$\mathtt{p}(|\psi\rangle) := \mathrm{tr}(P_{|\psi\rangle} P_1^{(n)}) = \| P_1^{(n)} |\psi\rangle \|^2.$$

From an intuitive point of view, the number $\mathtt{p}(\rho)$ represents the probability that the piece of quantum information ρ is true. We will see how this number will play an important role in our quantum computational semantics.

As is well-known, the basic idea of the theory of quantum computers is that computations can be performed by some quantum systems that evolve in time. Recalling that (according to Schrödinger's equation) the time-evolution of quantum systems is mathematically described by *unitary operators*, it is natural to assume that quantum information is processed by *quantum logical gates* (briefly, *gates*): special examples of unitary operators that transform in a reversible way the pure states of the quantum systems that store the information in question. All gates can be then extended to mixed states in a canonical way. On this basis, the computations performed by quantum computers can be mathematically described by *quantum circuits*: finite sequences $(\mathtt{G}_1^{(n)}, \ldots, \mathtt{G}_t^{(n)})$ of gates, all defined on a particular quantum computational space $\mathcal{H}^{(n)}$.

We will now introduce three important gates that have a special logical interest: the *negation-gate*, the *Toffoli-gate* and the *Hadamard-gate*.

Definition 9.11 (The Negation-Gate on the Space $\mathcal{H}^{(1)}$) The *negation-gate* on $\mathcal{H}^{(1)}$ is the linear operator $\mathtt{NOT}^{(1)}$ that satisfies the following condition for every element $|x\rangle$ of the canonical basis:

$$\mathtt{NOT}^{(1)}|x\rangle = |1 - x\rangle.$$

As happens in classical semantics, the quantum negation transforms the truth-value *Falsity* into the truth-value *Truth*, and vice versa. This gate can be naturally generalized to higher-dimensional spaces. For any $\mathcal{H}^{(n)}$ (with $n > 1$), the operator $\mathtt{NOT}^{(n)}$ is defined for every element $|x_1, \ldots, x_n\rangle$ of the canonical basis of the space as follows:

$$\mathtt{NOT}^{(n)}|x_1, \ldots, x_n\rangle := |x_1, \ldots, x_{n-1}\rangle \otimes \mathtt{NOT}^{(1)}|x_n\rangle.$$

Thus, the negation-gate always acts on the last bit of the input-register.

The smallest space where the Toffoli-gate can be defined is the space $\mathcal{H}^{(3)}$.

Definition 9.12 (The Toffoli-Gate on the Space $\mathcal{H}^{(3)}$) The *Toffoli-gate* on $\mathcal{H}^{(3)}$ is the linear operator $\mathtt{T}^{(1,1,1)}$ that satisfies the following condition for every element $|x, y, z\rangle$ of the canonical basis:

$$\mathtt{T}^{(1,1,1,)}|x, y, z\rangle := \begin{cases} |x, y, x \sqcap y\rangle, & \text{if } z = 0; \\ |x, y, (x \sqcap y)'\rangle, & \text{if } z = 1. \end{cases}$$

(Where $\sqcap$ and $'$ are the *infimum* and the *complement* of the two-valued Boolean algebra based on the set $\{0, 1\}$).

Thus, the Toffoli-gate leaves unchanged the first two bits $|x\rangle$ and $|y\rangle$ (which represent the *control-bits*); while the third bit $|z\rangle$ (representing the *target-bit*) is transformed into

- the bit corresponding to the Boolean conjunction of the two control-bits, when $z = 0$;
- the bit corresponding to the Boolean negation of the conjunction of the two control-bits, when $z = 1$.

Like the negation-gate, the Toffoli-gate also can be generalized to higher-dimensional spaces. For any $m, n \geq 1$, the Toffoli-gate $\mathtt{T}^{(m,n,1)}$ on the space $\mathcal{H}^{(m+n+1)}$ is defined as follows:

$$\mathtt{T}^{(m,n,1)}|x_1, \ldots, x_m, y_1, \ldots, y_n, z\rangle :=$$

$$|x_1, \ldots, x_{m-1}, y_{n-1}, y_1, \ldots, y_{n-2}\rangle \otimes \mathtt{T}^{(1,1,1)}|x_m, y_n, z\rangle.$$

The Toffoli-gate allows us to define a quantum-logical *conjunction* that behaves as a *reversible operation*. For any choice of two natural numbers m, n (such that $m, n \geq 1$) the reversible conjunction $\mathtt{AND}^{(m,n)}$ is dealt with as a *holistic* monadic operator that acts on *global* pieces of quantum information represented by quregisters of the space $\mathcal{H}^{(m+n)}$.

Definition 9.13 (The Conjunction on the Space $\mathcal{H}^{(m+n)}$) For any quregister $|\psi\rangle$ of $\mathcal{H}^{(m+n)}$

$$\mathtt{AND}^{(m,n)}|\psi\rangle := \mathtt{T}^{(m,n,1)}(|\psi\rangle \otimes |0\rangle)$$

(where the bit $|0\rangle$ plays the role of an *ancilla*).

As a special case consider a register $|x, y\rangle$ of the space $\mathcal{H}^{(2)}$. We obtain:

- $\mathtt{AND}^{(1,1)}|x, y\rangle = \mathtt{T}^{(1,1,1)}|x, y, 0\rangle = |1, 1, 1\rangle$, if $x = y = 1$.
- $\mathtt{AND}^{(1,1)}|x, y\rangle = \mathtt{T}^{(1,1,1)}|x, y, 0\rangle = |x, y, 0\rangle$, if $x = 0$ or $y = 0$.

Thus, $\mathtt{AND}^{(1,1)}$ represents a "good" quantum generalization of classical conjunction.

The negation-gate and the Toffoli-gate are examples of "semi-classical gates" that are unable to "create" superpositions from register-inputs. Of course, quantum-computer theory cannot help using also "genuine quantum gates" that can transform classical inputs (registers) into genuine superpositions, which are responsible for the characteristic parallel structures of quantum computations. An important example is represented by the *Hadamard-gate*.

Definition 9.14 (The Hadamard-Gate on the Space $\mathcal{H}^{(1)}$) The *Hadamard-gate* on $\mathcal{H}^{(1)}$ is the linear operator $\mathtt{H}^{(1)}$ that satisfies the following conditions:

$$\mathtt{H}^{(1)}|0\rangle = \frac{1}{\sqrt{2}}|0\rangle + \frac{1}{\sqrt{2}}|1\rangle; \;\; \mathtt{H}^{(1)}|1\rangle = \frac{1}{\sqrt{2}}|0\rangle - \frac{1}{\sqrt{2}}|1\rangle.$$

Thus, the Hadamard-gate transforms both bits into two distinct genuine superpositions that might be either *true* or *false* with probability $\frac{1}{2}$. Like the negation-gate, the Hadamard-gate also can be generalized to higher-dimensional spaces. For any $\mathcal{H}^{(n)}$ (with $n > 1$), the operator $\mathtt{H}^{(n)}$ is defined for every element $|x_1, \ldots, x_n\rangle$ of the canonical basis as follows:

$$\mathtt{H}^{(n)}|x_1, \ldots, x_n\rangle := |x_1, \ldots, x_{n-1}\rangle \otimes \mathtt{H}^{(1)}|x_n\rangle.$$

The Hadamard-gate (which cannot have any counterpart in classical computation, where superpositions do not exist) plays a very important role in quantum computation: creating superpositions means creating parallel structures, which are mainly responsible for the great speed and efficiency of quantum computers.

We will now introduce a special language whose aim is describing systems of quantum datasets. Let $DSSy = \{(\mathcal{H}_i, St_i, St_i^+, St_i^-, St_i^?)\}_{1 \le i \le n}$ be a quantum dataset-system, based on a set $Conc = \{C_1, \ldots, C_n\}$ of concepts. A quantum computational language for $DSSy$ is a language $\mathcal{L}_{DSSy}$ whose alphabet contains:

1. two special sentential constants **t** and **f** that denote the *Truth* and the *Falsity*, respectively;
2. a finite set of monadic predicates $\mathbf{P}_1, \ldots, \mathbf{P}_r$, where $r \geqslant n$;

3. a finite set of individual names $\mathbf{a}_1, \ldots, \mathbf{a}_s$, with $s \geqslant t$, where t is the maximum of the cardinal numbers of the sets St_i of $DSSy$;
4. the following logical connectives:
   - the negation $\neg$, corresponding to the gate negation;
   - the Hadamard -connective $\Diamond_{\mathtt{H}}$, corresponding to the Hadamard-gate;
   - the Toffoli-connective $\mathtt{T}$, corresponding to the Toffoli-gate.

The logical connectives simulate, at a syntactical level, the behavior of the corresponding gates. While the negation and the Hadamard-gate are 1-ary connectives (which act on single formulas), the Toffoli-connective is a ternary connective: if α, β are formulas and $\mathbf{q}$ is a sentential constant, then $\mathtt{T}(\alpha, \beta, \mathbf{q})$ is a formula. On this basis, recalling the definition of the quantum computational conjunction $\mathtt{AND}^{(m,n)}$, a binary conjunction $\wedge$ can be defined in terms of the Toffoli-connective:

$$\alpha \wedge \beta := \mathtt{T}(\alpha, \beta, \mathbf{f})$$

(where the false formula $\mathbf{f}$ plays the role of a *syntactical ancilla*).

Any formula α of $\mathcal{L}_{DSSy}$ can be decomposed into its parts, giving rise to a syntactical configuration called the *syntactical tree* of α. As an example, consider the (contradictory) formula

$$\alpha = \mathbf{Pa} \wedge \neg\mathbf{Pa} = \mathtt{T}(\mathbf{Pa}, \neg\mathbf{Pa}, \mathbf{f})$$

(say, *Alice is pretty and Alice is not pretty*). The syntactical tree of α is the following sequence of *levels*, where each level is a particular sequence of subformulas of α:

$$Level_3^\alpha = (\mathbf{Pa}, \mathbf{Pa}, \mathbf{f})$$

$$Level_2^\alpha = (\mathbf{Pa}, \neg\mathbf{Pa}, \mathbf{f})$$

$$Level_1^\alpha = (\mathtt{T}(\mathbf{Pa}, \neg\mathbf{Pa}, \mathbf{f}))$$

Generally, the levels of the syntactical tree of a given formula α are determined in the following way:

- the bottom level $Level_1^\alpha$ is (α);
- the top level $Level_h^\alpha$ is the sequence of atomic formulas occurring in α;
- $Level_{i+1}$ (where $1 \leq i < h$) is obtained by dropping the *principal connective* in all molecular formulas occurring at $Level_i$ and by repeating all atomic formulas that occur at $Level_i$.

The number of levels of the syntactical tree of α is called the *height* of α.

Consider a formula α such that:

$$Level_h^\alpha = (at_1^\alpha, \ldots, at_k^\alpha),$$

where $at_1^\alpha, \ldots, at_k^\alpha$ are the atomic formulas occurring at the top-level of the syntactical tree of α. The number k represents the *atomic complexity* of α (briefly indicated by $At(\alpha)$). This number will play an important semantic role, since it determines the *semantic space* $\mathcal{H}^{(At(\alpha))}$, where all possible meanings of α shall live. Instead of $\mathcal{H}^{(At(\alpha))}$ we will briefly write: $\mathcal{H}^\alpha$. For instance, in the case of the formula $\alpha = \intercal(\mathbf{Pa}, \neg\mathbf{Pa}, \mathbf{f})$, we have: $At(\alpha) = 3$. Hence, $\mathcal{H}^\alpha = \mathcal{H}^{(3)}$.

Another important concept that gives rise to a fundamental link between the syntactical world and the Hilbert-space world is the notion of *gate tree* of a given formula. As an example, let us refer again to the formula

$$\alpha = \mathbf{Pa} \wedge \neg\mathbf{Pa} = \intercal(\mathbf{Pa}, \neg\mathbf{Pa}, \mathbf{f}).$$

We have seen how in the syntactical tree of α the second level has been obtained from the third level by repeating the first occurrence of **Pa**, by negating the second occurrence of **Pa** and by repeating **f**, while the first level has been obtained by applying the connective $\intercal$ to the sequence of formulas occurring at the second level. Accordingly, one can say that the syntactical tree of α uniquely determines the following sequence consisting of two gates, both defined on the semantic space of α:

$$\left(\mathtt{I}^{(1)} \otimes \mathtt{NOT}^{(1)} \otimes \mathtt{I}^{(1)},\ \ \mathtt{T}^{(1,1,1)}\right)$$

(where $\mathtt{I}^{(1)}$ is the identity-gate of the space $\mathcal{H}^{(1)}$, $\mathtt{NOT}^{(1)}$ is the negation-gate of the space $\mathcal{H}^{(1)}$, $\mathtt{T}^{(1,1,1)}$ is the $(1, 1, 1)$-Toffoli-gate of the space $\mathcal{H}^{(3)}$). Such a sequence is called the *gate tree* of α. This procedure can be naturally generalized to any formula α. The general form of the gate tree of α will be:

$$(\mathsf{G}^{(\alpha)}_{(h-1)}, \ldots, \mathsf{G}^{(\alpha)}_{(1)}),$$

where h is the height of α.

Let us now turn to the semantics for our language $\mathcal{L}_{DSSy}$, where the main concept will be the notion of $\mathcal{L}_{DSSy}$*-model*. To this aim, we will first introduce the concepts of

- *atomic valuation* and *valuation* of $\mathcal{L}_{DSSy}$;
- *concept-representation* in the language $\mathcal{L}_{DSSy}$;
- *interpretation of the names of* $\mathcal{L}_{DSSy}$ *in* $DSSy$;
- $\mathcal{L}_{DSSy}$*-frame*.

Definition 9.15 (Atomic Valuation of $\mathcal{L}_{DSSy}$) An *atomic valuation* of $\mathcal{L}_{DSSy}$ is a map $\mathtt{val}_0$ that assigns to every atomic formula α of $\mathcal{L}_{DSSy}$ a meaning $\mathtt{val}_0(\alpha)$ represented by a qubit (in the space $\mathcal{H}^{(1)}$), satisfying the following condition:

$$\mathtt{val}_0(\mathbf{f}) = |0\rangle \ \text{ and } \ \mathtt{val}_0(\mathbf{t}) = |1\rangle.$$

Thus, the false formula and the true formula shall receive the "right" meanings, represented by the bit $|0\rangle$ and by the bit $|1\rangle$, respectively.

Definition 9.16 (Valuation of $\mathcal{L}_{DSSy}$) A *valuation* of $\mathcal{L}_{DSSy}$, based on an atomic valuation $\mathtt{val}_0$, is a map $\mathtt{val}$ that assigns to each level of the syntactical tree of every formula α a meaning represented by a quregister living in the semantic space of α.
Let $Level_h^\alpha = (at_1, \ldots, at_k)$ and let $(\mathsf{G}^{(\alpha)}_{(h-1)}, \ldots, \mathsf{G}^{(\alpha)}_{(1)})$ be the gate tree of α. The following conditions are required:

1. $\mathtt{val}(Level_h^\alpha) = \mathtt{val}_0(at_1) \otimes \ldots \otimes \mathtt{val}_0(at_k) \in \mathcal{H}^\alpha$.
2. For each $Level_i^\alpha$ such that $1 \leq i < h$,

$$\mathtt{val}(Level_i^\alpha) = \mathsf{G}_i^\alpha(\mathtt{val}(Level_{i+1}^\alpha)) \in \mathcal{H}^\alpha.$$

Thus, any valuation preserves the *logical form* of all subformulas that occur at a given level of the syntactical tree of any formula α.

On this basis, one can naturally identify the meaning assigned by a valuation $\mathtt{val}$ to a formula α with the meaning that $\mathtt{val}$ assigns to the bottom level of the syntactical tree of α :

$$\mathtt{val}(\alpha) := \mathtt{val}(Level_1^\alpha).$$

It is worthwhile noticing that (unlike classical extensional semantics) quantum computational meanings reflect, at least to a certain extent, the logical complexity of the formulas under investigation. For, the dimension of the semantic space, where all possible meanings of a given formula live, increases with the increasing of the complexity of the formula in question. In this sense, quantum computational meanings seem to behave like the *intensional meanings* of traditional logics.

Definition 9.17 (Concept-Representation) A *concept-representation* for $DSSy$ in the language $\mathcal{L}_{DSSy}$ is an injective map

$$\mathtt{cr} : Conc \mapsto \{\mathbf{P}_1, \ldots, \mathbf{P}_r\}$$

(where $Conc$ is the set of concepts of $DSSy$).

Accordingly, $\mathtt{cr}(C_i)$ represents the predicate that expresses the concept C_i in the language $\mathcal{L}_{DSSy}$.

Definition 9.18 (Interpretation of the Names) An *interpretation of the names* of $\mathcal{L}_{DSSy}$ in $DSSy$ is a partial function $\mathtt{int}$ whose domain is the set of all pairs $(i, \mathbf{a})$, where i is the index of a concept $C_i \in Conc$, while $\mathbf{a}$ is a name of $\mathcal{L}_{DSSy}$. When $int(i, \mathbf{a})$ is defined, we have: $int(i, \mathbf{a}) \in St_i$.

Of course, the function $\mathtt{int}$ is supposed to be partial, because, from an intuitive point of view, $\mathtt{int}(i, \mathbf{a})$ is reasonably defined only if the name $\mathbf{a}$ denotes an object-state for which the concept C_i makes sense (thus, $\texttt{int}(i, \mathbf{a}) \in St_i$). Instead of $\texttt{int}(i, \mathbf{a})$ we will briefly write: $\texttt{int}_i(\mathbf{a})$.

Definition 9.19 ($\mathcal{L}_{DSSy}$-Frame) An $\mathcal{L}_{DSSy}$-*frame* is a pair

$$\mathcal{F}r = (\texttt{cr}, \texttt{int}),$$

where $\texttt{cr}$ is a concept-representation for $DSSy$ in the language $\mathcal{L}_{DSSy}$, while $\texttt{int}$ is an interpretation of the names of $\mathcal{L}_{DSSy}$ in $DSSy$.

From an intuitive point of view it is natural to distinguish between *defined* and *non-defined* formulas with respect to a given frame.

Definition 9.20 (Defined Formulas) Consider an $\mathcal{L}_{DSSy}$-frame $\mathcal{F}r = (\texttt{cr}, \texttt{int})$ (where $Conc$ is the set of concepts of $DSSy$).

1. The sentential constants **f** and **t** are defined with respect to $\mathcal{F}r$.
2. An atomic formula **Pa** is *defined* with respect to $\mathcal{F}r$ iff $\texttt{cr}^{-1}(\mathbf{P}) = C_k \in Conc$ and $\texttt{int}_k(\mathbf{a})$ is defined (hence, $\texttt{int}_k(\mathbf{a}) \in St_k$).
3. A molecular formula α is *defined* with respect to $\mathcal{F}r$ iff all its atomic subformulas are defined with respect to $\mathcal{F}r$.

On this basis, we can now define the notion of a $DSSy$-model.

Definition 9.21 ($DSSy$-Model) A $DSSy$-model is a pair

$$\mathcal{M}_{DSSy} = (\mathcal{F}r, \texttt{val}),$$

where $\mathcal{F}r = (\texttt{cr}, \texttt{int})$ is an $\mathcal{L}_{DSSy}$-frame and $\texttt{val}$ is a valuation of $\mathcal{L}_{DSSy}$, satisfying the following conditions.

1. Suppose that **Pa** is defined for the frame $\mathcal{F}r$. Thus, there is a concept $C_k \in Conc$ such that

   $$\texttt{cr}^{-1}(\mathbf{P}) = C_k \;\text{ and }\; \texttt{int}_k(\mathbf{a}) \in St_k.$$

   In such a case we have:

   - $\texttt{val}(\mathbf{Pa}) = |1\rangle$, if $\texttt{int}_k(\mathbf{a}) \in St_k^+$.
   - $\texttt{val}(\mathbf{Pa}) = |0\rangle$, if $\texttt{int}_k(\mathbf{a}) \in St_k^-$.
   - $\texttt{val}(\mathbf{Pa}) = \frac{1}{\sqrt{2}}|0\rangle + \frac{1}{\sqrt{2}}|1\rangle$, if $\texttt{int}_k(\mathbf{a}) \in St_k^?$.

2. Suppose that **Pa** is not defined for the frame $\mathcal{F}r$. In such a case, $\texttt{val}(\mathbf{Pa})$ is a genuine qubit, different from the two bits $|0\rangle$ and $|1\rangle$.

Notice that for any formula α, the meaning $\texttt{val}(\alpha)$ exists in any model, even if α is not defined with respect to the frame of the model in question.

The concepts of *truth* in a $DSSy$-model and of *logical consequence* with respect to a given $DSSy$ can be now defined in the following way.

Definition 9.22 (Truth) Let $\mathcal{M}_{DSSy} = (\mathcal{F}r, \mathtt{val})$ be a $DSSy$-model.

$$\models_{\mathcal{M}_{DSSy}} \alpha \ \ (\alpha \text{ is } \textit{true} \text{ in } \mathcal{M}_{DSSy}) \ \text{ iff } \ \mathtt{p}(\mathtt{val}(\alpha)) = 1.$$

In other words, a formula is true in a given model, when its meaning has probability-value 1.

Definition 9.23 (Logical Consequence with Respect to a Given $DSSy$**)** $\alpha \models_{DSSy} \beta$ (β is a *logical consequence* of α) iff for any model $\mathcal{M}_{DSSy} = (\mathcal{F}r, \mathtt{val})$,

$$\mathtt{p}(\mathtt{val}(\alpha)) \leq \mathtt{p}(\mathtt{val}(\beta)).$$

In this semantics *truth* and *logical consequence* are clearly probabilistic concepts, where *probability* means *quantum probability*, according to the Born-rule.

In order to "see" from an intuitive point of how this peculiar semantics works in the case of particular applications, it may be useful to illustrate two characteristic examples.

Example: "Felix is a cat and Felix is beautiful" (Pa $\wedge$ Qa)

Consider a model $\mathcal{M}_{DSSy} = (\mathcal{F}r, \mathtt{val})$ and suppose that:
$\mathtt{cr}^{-1}(\mathbf{P}) = C_i$, $\mathtt{cr}^{-1}(\mathbf{Q}) = C_j$, $\mathcal{H}_i \neq \mathcal{H}_j$, $\mathtt{int}_i(\mathbf{a}) \in St_i$, $\mathtt{int}_j(\mathbf{a}) \in St_j$.
Thus, $\mathtt{int}_i(\mathbf{a}) \neq \mathtt{int}_j(\mathbf{a})$.

Such a situation seems quite reasonable from an intuitive point of view. For, the concept "beautiful" is certainly more complex than the concept "cat". Suppose that:

$$\mathtt{val}(\mathbf{Pa}) = \mathtt{val}(\mathbf{Qa}) = |1\rangle.$$

Hence,
$\mathtt{val}(\mathbf{Pa} \wedge \mathbf{Qa}) = \mathtt{val}(\mathtt{T}(\mathbf{Pa}, \mathbf{Qa}, \mathbf{f})) = |1, 1, 1\rangle$,
$\mathtt{p}(\mathtt{val}(\mathbf{Pa} \wedge \mathbf{Qa})) = 1$, $\models_{\mathcal{M}_{DSSy}} \mathbf{Pa} \wedge \mathbf{Qa}$.

Interestingly enough, this example also shows how just the semantics can create "trans-dataset identities", as happens in the case of the "trans-world identities" that arise in the framework of Kripke-semantics.

Example: "Felix is a cat and Felix is not a star" (Pa $\wedge$ ¬Qa)

Consider a model $\mathcal{M}_{DSSy} = (\mathcal{F}r, \mathtt{val})$ and suppose that:
$\mathtt{cr}^{-1}(\mathbf{P}) = C_i$, $\mathtt{cr}^{-1}(\mathbf{Q}) = C_j$, $\mathcal{H}_i \neq \mathcal{H}_j$, $\mathtt{int}_i(\mathbf{a}) \in St_i^+$. Then,
$\mathtt{val}(\mathbf{Pa}) = |1\rangle$.

(continued)

Two cases are possible:

1. $\texttt{int}_j(\mathbf{a}) \notin St_j$ Thus, **Qa** is not defined with respect to the frame $\mathcal{Fr}$. Hence,
   $\texttt{val}(\mathbf{Qa}) \neq |0\rangle, |1\rangle$,
   $\texttt{val}(\neg\mathbf{Qa}) \neq |0\rangle, |1\rangle$,
   $\texttt{val}(\mathbf{Pa} \wedge \neg\mathbf{Qa}) \neq |0\rangle, |1\rangle$,
   $\texttt{p}(\texttt{val}(\mathbf{Pa} \wedge \neg\mathbf{Qa})) \neq 0, 1$.
2. $\texttt{int}_j(\mathbf{a}) \in St_j$. In such a case, we will naturally assume that: $\texttt{int}_j(\mathbf{a}) \in St_j^-$.
   Hence,
   $\texttt{val}(\mathbf{Qa}) = |0\rangle$, $\texttt{val}(\neg\mathbf{Qa}) = |1\rangle$, $\texttt{val}(\mathbf{Pa} \wedge \neg\mathbf{Qa}) = |1\rangle$,
   $\texttt{p}(\texttt{val}(\mathbf{Pa} \wedge \neg\mathbf{Qa})) = 1$, $\models_{\mathcal{M}_{DSSy}} \mathbf{Pa} \wedge \neg\mathbf{Qa}$.

Notice that both cases admit a reasonable justification from an intuitive point of view. In the first case we suppose that our cat Felix does not belong to the conceptual domain of the predicate "star". In other words, we suppose that the question "Is Felix a star?" cannot be reasonably asked. Hence, the sentence "Felix is not a star" is considered meaningless. Consequently, also the conjunction "Felix is a cat and Felix is not a star" is meaningless and cannot be either true or false.

In the second case we suppose that the question "Is Felix a star?" can be reasonably asked (as is often assumed in the framework of classical semantics). Of course, as expected, the answer to this question will be negative: $\texttt{int}_j(\mathbf{a}) \in St_j^-$. Consequently, we obtain: $\models_{\mathcal{M}_{DSSy}} \mathbf{Pa} \wedge \neg\mathbf{Qa}$.

9.4 Concluding Remarks

We have seen how a quantum approach to machine learning allows us to simulate the intuitive notion of *Gestalt* that plays an important role in human recognitions and classifications of concepts. This is obtained by means of the mathematical concept of *positive centroid* of a quantum dataset associated to a particular concept (say, *table, triangle, beautiful, ...*). As happens with the concept of *Gestalt*, a quantum positive centroid (mathematically represented as a special example of quantum state) generally *alludes* in a vague and ambiguous way to the concrete instances that the agent under consideration has met in his/her previous experience. One is dealing with a characteristic form of *quantum ambiguity* that does not have any counterpart in the classical approaches to machine learning. Apparently, using the quantum theoretic formalism allows us to better understand and investigate possible relationships between human and artificial intelligences.

In the second Part of our Chapter we have seen how the theory of quantum datasets can be logically described and developed in the framework of a special form of quantum logical semantics, called *quantum computational semantics*, where *meanings* of formulas are represented as *pieces of quantum information*, while logical connectives are interpreted as *quantum logical gates*, corresponding to possible computation-actions. Such semantics represents a natural logical tool that allows us to *reason* about systems of quantum datasets, where different concepts are considered at the same time. In this way, one can study the logical behavior of molecular formulas (like "Alice is pretty and Bob is not jealous"), while the *indeterminate instances* of quantum datasets can be more deeply investigated by using the whole spectrum of quantum probabilities.

Finally, we would like to recall some open problems that arise in this field.

- Different kinds of quantum classifiers have been studied in the machine-learning literature. An important example is represented by *Helstrom-classifiers*, which are based on the concept of *physical discrimination* between quantum states by means of a convenient observable.[10] It would be interesting to investigate the relationships between *fidelity-classifiers* (considered in this Chapter) and *Helstrom-classifiers*, both from a mathematical and from an empirical point of view.
- A number of empirical simulations and concrete examples (that arise in some "real" scientific experiments) have shown the advantages of quantum classifiers, which in most cases turn out to be more *accurate* with respect to classical classifiers.[11] However, the general theoretic reasons that are responsible for this form of "quantum advantage" are not clear and should require a deeper investigation. In particular, so far, the role of *entanglement* and of *non-locality* in the quantum approaches to machine learning has not been properly studied.
- From a logical point of view it might be interesting to develop a quantum computational semantics for systems of quantum datasets, where concepts representing binary (or, more generally, n-ary) relations are considered.

Acknowledgments R. Giuntini and G. Sergioli are partially supported by the projects: (1) "Ubiquitous Quantum Reality (UQR): understanding the natural processes under the light of quantum-like structures", funded by Fondazione di Sardegna (code: F73C22001360007); (2) "CORTEX The COst of Reasoning: Theory and EXperiments", funded by Ministry of University and Research (Prin 2022): (3) "Quantum Models for Logic, Computation and Natural Processes (Qm4Np)" funded by Ministry of University and Research (Prin-Pnrr 2022). R. Giuntini is partially funded by the TÜV SÜD Foundation, the Federal Ministry of Education and Research (BMBF) and the Free State of Bavaria under the Excellence Strategy of the Federal Government and the Länder, as well as by the Technical University of Munich-Institute for Advanced Study.

[10] See, for instance, Sergioli et al. (2017) and Sergioli et al. (2019).

[11] See, for instance, Sergioli et al. (2021).

References

Dalla Chiara, M.L., R. Giuntini, R. Leporini, and S. Sergioli, S. 2018. *Quantum Computation and Logic. How Quantum Computers Have Inspired Logical Investigations*. Berlin: Springer.

Ehrenstein, W.H., L. Spillmann, and W. Sarris. 2003. Gestalt Issues in Modern Neuroscience. *Axiomathes* 13: 433–458.

Giuntini, G., F. Holik, D.K. Park, H. Freytes, C. Blank, and G. Sergioli. 2023a. Quantum-Inspired Algorithm for Direct Multi-Class Classification. *Applied Soft Computing* 134: 109956.

Giuntini, R., A.C. Granda Arango, H. Freytes, F.H. Holik, and G. Sergioli. 2023b. Multi-Class Classification Based on Quantum State Discrimination. *Fuzzy Sets and Systems* 467: 108509.

Kofka, K. 1935. *Principles of Gestalt Psychology*. New York: Harcourt.

Köhler, W. 1920. *Die physischen Gestalten in Ruhe und in stationären Zustand*. Braunschweig: Vieweg.

Schuld, M., and F. Petruccione. 2018. *Supervised Learning with Quantum Computers*. Berlin: Springer.

Sergioli, G., G.M. Bosyk, E. Santucci, and R. Giuntini. 2017. A Quantum-Inspired Version of the Classification Problem. *International Journal of Theoretical Physics* 56: 3880–3888.

Sergioli, G., R. Giuntini, and G. Freytes. 2019. A New Quantum Approach to Binary Classification. *PLOS ONE* 14 (5): e0216224.

Sergioli, G., C. Militello, L. Rundo, L. Minafra, F. Torrisi, G. Russo, K.L. Chow, and R. Giuntini. 2021. A Quantum-Inspired Classifier for Clonogenic Assay Evaluations. *Scientific Reports* 11: 2830.

Wertheimer, M. 1912. Experimentelle Studien über das Sehen von Bewegung. *Zeitschrift für Psychologie* 61: 161–265.

Index

H. Hosni, J. Landes (eds.), *Perspectives on Logics for Data-driven Reasoning*,
Logic, Argumentation & Reasoning 35,
https://doi.org/10.1007/978-3-031-77892-6

MIX
Papier aus verantwortungsvollen Quellen
Paper from responsible sources
FSC® C105338

If you have any concerns about our products,
you can contact us on
ProductSafety@springernature.com

In case Publisher is established outside the EU,
the EU authorized representative is:
Springer Nature Customer Service Center GmbH
Europaplatz 3, 69115 Heidelberg, Germany

Printed by Libri Plureos GmbH
in Hamburg, Germany